Course Design for Machinery Design
机械设计课程设计
（英文版）

何　波　主编

西北工业大学出版社

【内容简介】 本书内容共分为两个部分。第一部分为机械设计课程设计指导,以常见的二级圆柱齿轮减速器为例,系统介绍机械传动装置的设计内容、步骤和方法。第二部分为机械设计课程设计常用标准和规范,列出机械设计课程设计的常用标准、规范和设计资料,并提供一些参考图例。

本书可作为高等院校机械类专业机械设计课程双语教学或全英文授课的教材或教学参考书。

图书在版编目(CIP)数据

机械设计课程设计＝Course Design for Machinery Design：英文/何波主编. —西安：西北工业大学出版社,2017.6
ISBN 978 - 7 - 5612 - 5384 - 7

Ⅰ.①机… Ⅱ.①何… Ⅲ.①机械设计—课程设计—高等学校—教材—英文 Ⅳ.①TH122 - 41

中国版本图书馆 CIP 数据核字(2017)第 124582 号

策划编辑：杜　莉
责任编辑：胡莉巾

出版发行：	西北工业大学出版社
通信地址：	西安市友谊西路 127 号　　邮编：710072
电　　话：	(029)88493844　88491757
网　　址：	www.nwpup.com
印　刷　者：	陕西天意印务有限责任公司
开　　本：	787 mm×1 092 mm　　1/16
印　　张：	11.125
字　　数：	249 千字
版　　次：	2017 年 6 月第 1 版　　2017 年 6 月第 1 次印刷
定　　价：	35.00 元

前　言

　　机械设计是机械类专业的一门重要技术基础课，课程设计是该课程必要的实践环节。《机械设计课程设计》（英文版）是根据高等工科院校机械类专业机械设计课程双语和全英文教学的需求，结合沈阳航空航天大学及兄弟院校在机械设计课程设计双语和全英文教学方面的经验编写而成的，与《机械设计》配套使用。

　　全书共分为两个部分。第一部分为机械设计课程设计指导，以常见的二级圆柱齿轮减速器为例，系统介绍机械传动装置的设计内容、步骤和方法。第二部分为机械设计课程设计常用标准和规范，列出机械设计课程设计的常用标准、规范和设计资料，并提供一些参考图例。全书共20章。第一部分分为8章：概述、传动装置的总体设计、传动零件的设计计算、减速器的构造、减速器装配草图设计、零件工作图设计、装配工作图设计、设计计算说明书编写。第二部分分为12章：机械制图、常用资料和标准、螺纹连接、键和销连接、滚动轴承、联轴器、润滑与密封、减速器附件、圆柱齿轮结构、公差配合和表面粗糙度、电动机以及参考图例和设计实例。

　　在本书编写过程中，得到了东北大学巩云鹏教授的指导和支持，在此谨致谢意！

　　衷心希望广大读者对书中不妥之处提出宝贵意见。

何　波
2017年1月

Preface

Machinery Design is an important technological basic course in mechanical engineering education and the course design is an essential practice. This English textbook *Course Design for Machinery Design* is written for Chinese and international students majoring in mechanical engineering or in other related field to meet the basic requirements for this course in advanced engineering universities and colleges. And teaching experiences of Shenyang Aerospace University and other universities in the Course Design for Machinery Design have been combined. This textbook should be used together with *Machinery Design* and many Chinese textbooks have been referred to during the compilation.

There are two parts in this textbook: instruction of course design for machinery design and design information and data. In the first part, the design contents, procedure and method for a two-level cylindrical gear reducer have been introduced. In the second part, the commonly used standards, regulations and design data have been illustrated and some examples have been provided. There are 20 chapters in this textbook. The first part includes 8 chapters: overview, outline design of transmission equipments, design calculations for transmission components, construction of reducer, assembly sketch design, component drawing, assembly drawing design, design and calculation description. The second part includes 12 chapters: drawing, commonly used data and general standard, thread fasteners, key and pin, rolling-contact bearing, coupling, lubrication and sealing, associated component, structure of cylindrical gear, limit and fit, tolerance and surface roughness, electromotor, drawing examples and cases.

Professor Gong Yunpeng from Northeastern University has kindly directed the compilation and put forward some insightful suggestions.

Since it is our first attempt to compile a technical textbook in English, there would be some mistakes and inappropriate treatments inevitably in the book. So your suggestions and criticism are welcomed sincerely.

He Bo
2017. 1

Contents

PART I INSTRUCTION OF COURSE DESIGN FOR MACHINERY DESIGN

Chapter 1 Overview 3

    1.1 Purposes of Machinery Design Course Design 3

    1.2 Contents of Machinery Design Course Design 3

    1.3 Procedure and Schedule of Machinery Design Course Design 4

    1.4 Approaches to Machinery Design Course Design and Notes 4

Chapter 2 Outline Design of Transmission Equipments 6

    2.1 Determination of Transmission Scheme 6

    2.2 Types of Reducer 7

    2.3 Selection of Motor 8

    2.4 Distribution of Speed Ratio 11

    2.5 Calculation of Kinetic and Dynamic Parameters 13

Chapter 3 Design Calculations for Transmission Components 16

    3.1 Design Calculation for Transmission Components out of the Reducer 16

    3.2 Design Calculation for Transmission Components in the Reducer (Gear Design) 18

Chapter 4 Construction of Reducer 20

    4.1 Gear, Shaft and Bearing Combination 21

    4.2 Housing 21

    4.3 Associated Components 22

Chapter 5 Assembly Sketch Design 26

    5.1 Tentative Drawing of Assembly Sketch 26

    5.2 Strength Calculation of Shafts, Bearings and Keys 34

    5.3   Completion of Assembly Sketch Design ·················· 35

Chapter 6   Component Drawing ································ 40

    6.1   Design Requirements of Component Drawing ·········· 40

    6.2   Shaft Drawing Design ······································ 42

    6.3   Gear Drawing Design ······································· 43

    6.4   Housing Drawing Design ···································· 44

Chapter 7   Assembly Drawing Design ························· 46

    7.1   Views ··· 46

    7.2   Dimensioning ·· 46

    7.3   Component No., Title Block and Component List ········ 47

    7.4   Technical Feature of Reducer ····························· 48

    7.5   Technical Condition ··· 49

    7.6   Assembly Drawing Check ·································· 51

    7.7   Error Correction Practice for Assembly Drawing ········ 51

Chapter 8   Design and Calculation Description ············· 55

    8.1   Contents and Requirements ································ 55

    8.2   Outline ··· 56

PART Ⅱ   DESIGN INFORMATION AND DATA

Chapter 9   Drawing ··· 59

    9.1   General Regulations ·· 59

    9.2   Drawing of Commonly Used Components ················ 61

Chapter 10   Commonly Used Data and General Standard ···· 63

    10.1   Commonly Used Data ····································· 63

    10.2   General Standard ·· 64

Chapter 11   Thread Fasteners ································ 67

    11.1   Thread Hole ·· 67

    11.2   Bolt ·· 68

    11.3   Screw ··· 71

    11.4   Nut ··· 75

    11.5   Washer ··· 76

    11.6   Fastening End Cap ·· 77

Chapter 12　Key and Pin ………………………………………………………… 78

　12.1　Key ……………………………………………………………………… 78

　12.2　Pin ……………………………………………………………………… 80

Chapter 13　Rolling-contact Bearing …………………………………………… 81

Chapter 14　Coupling ……………………………………………………………… 88

Chapter 15　Lubrication and Sealing …………………………………………… 94

　15.1　Lubricant ……………………………………………………………… 94

　15.2　Oil Indicator ………………………………………………………… 96

　15.3　Sealing ………………………………………………………………… 96

Chapter 16　Associated Component ……………………………………………… 99

　16.1　Checking Hole and Checking Hole lid …………………………… 99

　16.2　Ventilator …………………………………………………………… 99

　16.3　Bearing Cover ……………………………………………………… 100

　16.4　Oil Baffle …………………………………………………………… 101

　16.5　Lifting Equipment ………………………………………………… 102

Chapter 17　Structure of Cylindrical Gear …………………………………… 103

Chapter 18　Limit and Fit, Tolerance and Surface Roughness …………… 105

　18.1　Limit and Fit ………………………………………………………… 105

　18.2　Standard Tolerance and Limit Deviation ……………………… 106

　18.3　Surface Roughness ………………………………………………… 111

　18.4　Accuracy Degree of Gear ………………………………………… 114

Chapter 19　Electromotor ……………………………………………………… 115

Chapter 20　Drawing Examples and Cases …………………………………… 118

　20.1　Drawing Examples ………………………………………………… 118

　20.2　Case 1 ………………………………………………………………… 126

　20.3　Case 2 ………………………………………………………………… 150

Reference …………………………………………………………………………… 168

PART Ⅰ INSTRUCTION OF COURSE DESIGN FOR MACHINERY DESIGN

Chapter 1 Overview

1.1 Purposes of Machinery Design Course Design

"Machinery design course design" is one of the important practice steps in machinery design. The course design has significant meanings:
(1) To enhance students' ability to link machinery design theory with practice.
(2) To learn the general approach to machinery design.
(3) To make a training of the basic skills on machinery design, including: calculation, drawing, and usage of the design data, manuals, standards and specifications.
(4) To have a correct design thought. Not only the innovation opinion, but also the existent experience is very important.

Besides, machinery design course design also establishes the foundation for specialized course design and graduation design.

1.2 Contents of Machinery Design Course Design

In machinery design course design, the mechanical transmission device of a general machinery is usually selected as the design subject. As shown in Fig. 1 - 1, the transmission system includes gear, worm gear, belt, chain, coupling and other parts, and general knowledge of the mechanical design is involved. Reducer design is suitable for the students' knowledge level, and can help students to get a basic comprehensive training.

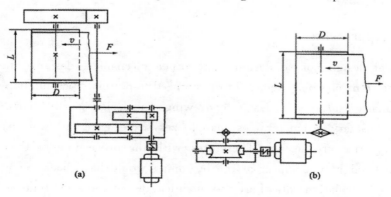

Fig. 1 - 1 Types of design titles

The contents of the course design include:
(1) Determining the transmission scheme;
(2) Choosing motor, calculating the total transmission ratio and distributing the ratios at all levels, calculating the motion and kinetic parameters;
(3) Designing the transmission parts;
(4) Designing shafts, selecting and checking key joints;
(5) Designing bearing combinations;
(6) Designing housing and accessories;
(7) Designing lubrication and sealing systems;
(8) Designing and drawing assembly drawing and component drawings;
(9) Writing the design and calculation description.

1.3　Procedure and Schedule of Machinery Design Course Design

(1) Preparation: to understand the design task, requirements, working conditions and contents;
(2) Outline design of the transmission equipment: to select the type of motor, to calculate the kinetic and dynamic parameters of the transmission equipment;
(3) Design calculation for transmission components;
(4) Assembly sketch design;
(5) Working drawing design of components and assembly;
(6) Description writing for design and calculation.

1.4　Approaches to Machinery Design Course Design and Notes

1.4.1　Approaches

Mechanical design course design, like other mechanical design, starts from the analysis of the transmission scheme. Then you need to do related calculation and structure design, and finally express your design by drawings and design specifications. In the design process, some primary parameters, estimated sizes, experience data were used during the calculation and structure design. Therefore, with the proceeding of the design, some problems which might not appear at the beginning would have been exposed gradually, which requires "calculation, drawing, and modification" at the same time.

Any attempt to determine all the sizes and structures of the parts only by the method of theoretical calculation without sketch drawing, or the reluctance to make necessary changes once the sketch is drawn makes it incorrect.

1.4.2 Notes

(1) It is necessary to do design work seriously, carefully, and neatly. Any slight mistake in calculation or structure design may lead to product scrap.

(2.) To make your design reasonable, practical, economical and of good manufacturability, comprehensive thinking and combining theory with practice are important.

(3) It is necessary to correctly realize the relationship between inheritance and innovation, and use standards and specifications properly. Correct usage of the standard specification is not only favorable to the interchangeability and manufacturability of parts, which leads to good economic benefits, but also it can reduce the workload and accelerate the design. Generally, national standard and the department specification should be strictly followed. However, standards and specifications are not made to limit innovation and development. Therefore, when standards and specifications have contradictions with the design requirements, you should design them by yourself.

(4) It is necessary to deal well with relations between calculation and structure design, and make overall decision after taking all factors into consideration.

1) Sizes calculated through the formula derived from geometric relationship are very strict. Generally, they cannot be rounded off or changed. For example, the center distance of gear transmission $a = \dfrac{m(z_1 + z_2)}{2}$. If a is rounded off, z_1, z_2 or m should be changed correspondingly, in order to keep identity relation.

2) Formula derived from the strength, stiffness and wear conditions are usually in inequality relationship. For example, according to the strength calculation, diameter of a shaft is at least 32 mm, but the size used finally may be 50 mm, considering the structure, installation, disassembly, processing and manufacturing requirements of other parts that the shaft fits with (such as coupling, gear and rolling bearing, etc.).

3) Empirical formulas concluded by the practice, are often used to determine sizes, when the shape is complex, or the strength is uncertain, such as the housing. And the sizes are often rounded off.

4) Secondary sizes can be determined by the designers themselves, the processing and using conditions should be considered and similar structure can be referred to, such as fixed bushing and oil baffle plate, etc.

(5) Drawing should be expressed correctly and clearly, conforming with the standard of mechanical drawing; design and calculation description should be accurate and neat, conforming with the requirements of the writing format.

Chapter 2　Outline Design of Transmission Equipments

The tasks of the transmission's outline design include: determining the transmission scheme, selecting the type of motor, distributing transmission ratio reasonably and calculating the kinetic and dynamic parameters of the transmission equipment. It can prepare conditions for the design and calculation of all levels of the transmission parts.

2.1　Determination of Transmission Scheme

Reasonable transmission scheme should meet the performance requirements of the working part and work reliably. It should also have the advantage of simple structure, compact size, easy processing, low cost, high efficiency and convenience in operation and maintenance etc. But it is very difficult to meet all the requirements at the same time. So we should balance them and ensure the key requirements.

When multi-stage transmission is used, we should reasonably select transmission parts and determine the transmission order of them, to gain a reasonable scheme. The following points should often be considered.

(1) Belt drive is a kind of friction drives. Its transmission is steady and it can absorb quite a large amount of sudden shock. But its transmission ratio is neither constant nor exact. And while transmitting the same torque, the structure size of the belt drive is bigger than that of other form of transmission. So belt drives should be placed as high-speed transmission. While transmitting the same power, if the rotation speed is higher, the torque would be smaller.

(2) The operation of chain drive depends on the meshing of chain and sprockets, and its average transmission ratio can keep constant. Moreover, chain drive can operate under an adverse working environment. But the instantaneous velocity of the chain is not constant, leading to impacts. So, generally, chain drives are used at low-speed levels

instead of high-speed ones.

(3) The open gearing should be arranged at low-speed levels because of its poor working condition, bad lubrication and severe wear.

(4) Usually, helical gear is used in multi-stage gear drive, because its stability is better than spur gear.

2.2 Types of Reducer

The reducer is usually used as an individual unit to transmit power between motor and working part. Reducers have some features, such as compact structure, high transmission efficiency, reliable drive and convenient maintenance, which makes them extensively employed in various machines.

In order to accommodate different purposes for all kinds of mechanical transmission, reducers have many styles, whose features and applications are shown in Table 2-1.

Table 2-1 Features and applications of reducers

Name	Kinematic diagram	Recommended range of speed ratio	Features and applications
Single cylindrical gear reducer		$i \leqslant 10$	There are spur gears, helical gears and herringbone gears. Spur gears are applied in the transmission whose speed is low or workload is small. For higher speed and heavier transmission load, helical gears or herringbone gears should be adopted. Most housings are made of cast iron, and sometimes welded structure or casting steel is used. Rolling-contact bearings are normally employed, and only under heavy load or specially high speed transmission, sliding bearings are used. Other types are similar to the above

Continued

Name	Kinematic diagram	Recommended range of speed ratio	Features and applications
Two cylindrical gear reducer — Expanded type		$i = 8 - 60$	The structure of expanded type is simple, but gear's position relating to the bearing is not symmetric, so the shaft should possess greater stiffness. In order to decrease the nonuniformity of load distribution along the gear width, high-speed stage gear should be located far away from the input of the torque. It is recommended that this type is adapted to stable load. Helical gears can be applied to high-speed stage or low-speed stage, but spur gears are better to be utilized in low one
Two cylindrical gear reducer — Coaxial type		$i = 8 - 60$	The length of this reducer is short. The immersion depths into oil are approximately equal for the two pairs of the gears. But the axial size and weight are larger, and the carrying capacity of high-speed gears cannot make full use; Load is not uniform along the gear width, for the middle shaft is too long and the stiffness is decreased. There is only one shaft end for input and output, which limits the flexibility of layout

2.3 Selection of Motor

According to magnitude and properties of workload, characteristics of working parts and working conditions, the category, type, structure form, power and speed for the motor can be selected, so the model of motor will be determined.

2.3.1 Category, Type and Structure Form of Motor

The category, type and structure forms can be selected based on power type (DC or AC), working conditions (environment, temperature and position), magnitude and properties of workload, starting characteristics, overload situation, etc.

Chapter 2 Outline Design of Transmission Equipments

Usually without special requirements, three-phase AC motors are adopted, especially the Y series motors. If the system starts, brakes, inverts frequently, such as hoisting equipment, the motor is required to have small rotational inertia and large overload capacity, so three-phase asynchronous motor for metallurgy and hoisting could be chosen, such as YZ series (squirrel-cage) or YZR series (wound-rotor). The technical data and outline dimensions of motors are shown in Table 19-1 and Table 19-2.

2.3.2 Power of Motor

The operation and economy for motor are both affected by the power selection. Too low power can't sustain the normal work of the machine, leading to the motor failures in advance due to overload. And too high power can also bring many drawbacks, such as high price, inability to fully play, and lower efficiency and power factor, and waste will be made.

For the machine whose workload is stable (or changes very little) and operation is continuous for a long time (as conveyer), the motor can be selected according to the rated power, and the verifying of heat and starting torque of the motor is not required. The following should be insured:

$$P_0 \geqslant P_r \tag{2-1}$$

where P_0— rated power of motor (kW);

P_r— required power of working part (kW).

P_r can be calculated as below:

$$P_r = \frac{P_W}{\eta} = \frac{Fv}{1,000\eta} \tag{2-2}$$

where P_W— required effective power of working part (kW);

η — total efficiency from motor to working part;

F — peripheral force of working part (N);

v — linear velocity of working part (m/s).

For different special machine, the calculation of P_W is different. For example, belt conveyor:

$$P_W = \frac{Fv}{1,000} = \frac{F\pi nD}{60 \times 10^6} \tag{2-3}$$

where D — diameter of driving roller of belt conveyor (mm);

n — speed of driving shaft of working part (r/min).

The total efficiency η is determined by the composition of transmission equipments as follows:

$$\eta = \eta_1 \eta_2 \eta_3 \cdots \eta_W$$

where $\eta_1, \eta_2, \eta_3, \ldots, \eta_W$ are the efficiencies of kinematic pairs or transmission pairs (such as coupling, gear, belt, bearing and roller).

In the calculation of the total efficiency, the following points should be taken into account:

(1) The determination of efficiency of kinematic pair or transmission pair can refer to Table 10-1 where the ranges of efficiency are provided. If the situation is ambiguous, the middle value can be taken. If the working conditions are bad, the machining accuracy is low, and the maintenance is poor, a low value should be taken, and vice verse.

(2) For each kinematic pair or transmission pair, the power will be lost more or less. So pay attention not to omit anyone.

(3) The efficiency of bearing is for a pair of bearings.

2.3.3 Determination of Rotating Speed of Motor

After the capacity and type have been selected, the rotating speed should be determined. For instance, the synchronous speed of three-phase asynchronous motor has four values (3,000 r/min, 1,500 r/min, 1,000 r/min and 750 r/min). High-speed motor has the following advantages: less pole, simple structure, small outline dimensions and low price. But too high speed results in bigger speed ratio which leads to complicating the structure, increasing the outline dimensions and raising the manufacturing cost. So analysis and comparison will help to select an optimum scheme. In this course design, the suggested synchronous speed is 1,500 r/min or 1,000 r/min, and 750 r/min is only used in some special situation.

Example 2-1

There is a belt conveyor as shown in Fig. 2-1. Its peripheral force (F), linear velocity (v) and diameter of roller are respectively 6,000 N, 0.5 m/s and 300 mm. The load is stable, the operation is continuous, working condition is dusty, and the power is three-phase AC whose voltage is 380 V. Please select a suitable motor.

Solution

(1) Choose the type.

Three-phase asynchronous motor was chosen whose structure is closed and voltage is 380V, according to the requirement. The Y series motors are adopted.

(2) Select the power.

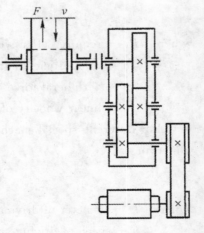

Fig. 2-1 A belt conveyor

The required effective power of the roller is

$$P_w = \frac{Fv}{1,000} = \frac{6,000 \times 0.5}{1,000} = 3.0 \text{ kW}$$

Total efficiency is $\eta = \eta_{belt} \cdot \eta_{gear}^2 \cdot \eta_{bearing}^4 \cdot \eta_{coupling} \cdot \eta_{rollor}$.

According to Table 10-1,

Chapter 2　Outline Design of Transmission Equipments

$$\eta_{\text{belt}} = 0.96$$
$$\eta_{\text{gear}} = 0.97 \text{ (class 8)}$$
$$\eta_{\text{bearing}} = 0.99 \text{ (ball bearing)}$$
$$\eta_{\text{coupling}} = 0.99$$
$$\eta_{\text{rollor}} = 0.96$$

So
$$\eta = 0.96 \times 0.97^2 \times 0.99^4 \times 0.99 \times 0.96 = 0.825$$

Required power is
$$P_r = P_W/\eta = 3.0/0.825 = 3.64 \text{ kW}$$

According to Table 19-1, choose Y112M-4 whose rated power is 4 kW or Y132MI-6 whose rated power is also 4 kW.

(3) Determine the rotating speed.

The rotating speed of the roller is
$$n_W = \frac{60v}{\pi D} = \frac{60 \times 0.5}{\pi \times 0.3} = 31.8 \text{ r/min}$$

There are two plans whose synchronous speeds are respectively 1,500 r/min and 1,000 r/min as shown in Table 2-2, according to Table 19-1.

Table 2-2　Motor parameters and total speed ratio for motor

Plan No.	Model of motor	Rated power/kW	Synchronous speed/(r·min^{-1})	Speed with full load/ (r·min^{-1})	Quality of motor/kg	Total speed ratio
1	Y112M-4	4.0	1,500	1,440	51	45.28
2	Y132M1-6	4.0	1,000	960	73	30.19

Comparing these two plans, No. 1 has light weight and low price, but its total speed ratio is greater. To compact the structure, we choose No. 2. From Table 19-2, the height of the center (H) is 132 mm, and the overhang shaft ($D \times E$) is 38 mm$\times$80 mm.

2.4　Distribution of Speed Ratio

According to the motor speed with full load (n_0) and the speed of working part (n_W), the total speed ratio of transmission can be calculated by $i = n_0/n_W$. Then i should be distributed reasonably to each level. It can be expressed as $i = i_1 i_2 \ldots i_n$.

To distribute i reasonably, the following points should be taken into account:

(1) Speed ratio of each level should be in rational ranges to compact structure and rationalize process. The recommended speed ratios are shown in Table 10-1.

(2) The size and structure of different level should be considered and determined rationally. For example, as shown in Fig. 2-2, the speed ratio of belt drive should not be too large and should satisfy $i_{belt} < i_{gear}$. If the speed ratio of the belt is too large, the radius of the big pulley will be greater than the height of the center, which causes difficult installation.

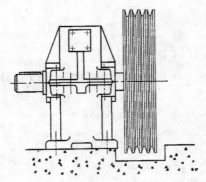

Fig. 2-2 Difficult installation caused by too big pulley

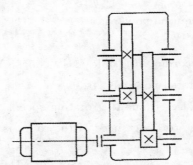

Fig. 2-3 Interference between large gear from high-speed level and low-speed shaft

(3) Each transmission part should not interference with each other. For example, if the speed ratio of high-speed level for two-level cylindrical gear reducer is too big, a collision may occur between the tip circle of the large gear from high-speed level and the shaft of the large gear from low-speed level, as shown in Fig. 2-3. If i is distributed inappropriately, there may be an interference between roller and the shaft of the small gear from the open gearing.

(4) When spread two-level gear reducer is designed and bath lubrication is used, the oiled depth of the larger gears of high speed level and low speed level should be close (see Fig. 2-4). When the mating materials of the two levels are the same and the tooth face width factors are same, the contact fatigue strengths should be made approximately equal, the distribution of speed ratio can be done according to

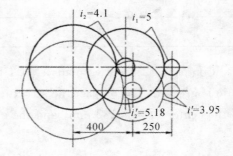

Fig. 2-4 Influence of speed ratio on reducer

$$i_1 = (1.3 - 1.4) i_2, \quad i_1 = \sqrt{(1.3 - 1.4) i_{reducer}}$$

where i_1, i_2—the speed ratio of high-speed level and low-speed level respectively;

$i_{reducer}$—the speed ratio of the reducer.

Example 2-2

The data are same with Example 2-1. The speed with full load (n_0) is 960 r/min, and total speed ratio is $i = \dfrac{n_0}{n_W} = \dfrac{960}{31.8} = 30.19$. Please distribute the speed ratio.

Solution

According to Table 10-1, take 2.5 as the speed ratio of the belt, so the speed ratio of reducer is

$$i_{\text{reducer}} = \frac{i}{i_{\text{belt}}} = \frac{30.19}{2.5} = 12.076$$

The speed ratio of high-speed level for two-level gear reducer is

$$i_1 = \sqrt{1.35 i_{\text{reducer}}} = \sqrt{1.35 \times 12.076} = 4.038$$

So the speed ratio of low-speed level is

$$i_2 = i_{\text{reducer}}/i_1 = 12.076/4.038 = 2.989$$

Note that the above distribution is primary. The actual speed ratio should be calculated after the determination of parameters of each transmission. It is allowed that the actual speed ratio does not comply with the required one, and the error should be within $\pm(3-5)\%$.

2.5 Calculation of Kinetic and Dynamic Parameters

After the determination of motor and distribution of speed ratio, the transmitted power, rotating speed and torque of each shaft can be calculated, providing basis for the design and calculation of transmission components and shafts.

(1) According to speed with full load of the motor and speed ratios, rotating speed of each shaft can be calculated.

(2) The power and torque of each shaft should be calculated according to the input. There are two calculating methods. In one method, the required power of working part is used. In the other method, the rated power of the motor is used. The former has the advantage that the structure of transmission is more compact. For the latter, the carrying capacity of transmission has some potential because the rated power (P_0) is bigger than the required power (P_r).

Each shaft from high speed to low speed is defined as shaft Ⅰ, shaft Ⅱ ... respectively. The speed ratio and efficiency of two adjacent shafts are respectively i_{12}, i_{23}... and η_{12}, η_{23}.... Input power of each shaft is P_1, P_2... Rotating speed is n_1, n_2... Input torque is T_1, T_2... So the calculation formulae are as follows:

$$\left. \begin{array}{llll} P_0 & n_0 & T_0 = 9.55 \dfrac{P_0}{n_0} & \\[6pt] P_1 = P_0 \cdot \eta_{01}, & n_1 = \dfrac{n_0}{i_{01}}, & T_1 = 9.55 \dfrac{P_1}{n_1} = T_0 \cdot i_{01} \cdot \eta_{01} & \\[6pt] P_2 = P_1 \cdot \eta_{12}, & n_2 = \dfrac{n_1}{i_{12}}, & T_2 = 9.55 \dfrac{P_2}{n_2} = T_1 \cdot i_{12} \cdot \eta_{12} & \\[6pt] P_3 = P_2 \cdot \eta_{23}, & n_3 = \dfrac{n_2}{i_{23}}, & T_3 = 9.55 \dfrac{P_3}{n_3} = T_2 \cdot i_{23} \cdot \eta_{23} & \end{array} \right\} \quad (2-4)$$

where P_0 — output power of motor (kW).

n_0 — speed with full load of motor (r/min).

T_0 — output torque of motor (N·m).

i_{01} — the speed ratio from motor shaft to shaft I; if they are connected by a coupling, then $i_{01}=1$.

η_{01} — the efficiency from motor shaft to shaft I.

If the first method is used, P_0 is the required power, that is, $P_0 = P_r$. If the second method is used, P_0 is the rated power.

Example 2-3

The data are the same with Example 2-1. Please calculate the kinetic and dynamic parameter of each shaft. The first method is used in the example.

Solution

Shaft 0:
$$P_0 = P_r = 3.64 \text{ kW}$$
$$n_0 = 960 \text{ r/min}$$
$$T_0 = \frac{9.55 P_0}{n_0} = 9.55 \times 3.64 \times 10^3 / 960 = 36.21 \text{ N·m}$$

Shaft I:
$$P_1 = P_0 \eta_{01} = P_0 \eta_{belt} = 3.64 \times 0.96 = 3.49 \text{ kW}$$
$$n_1 = \frac{n_0}{i_{01}} = \frac{n_0}{i_{belt}} = \frac{960}{2.5} = 384 \text{ r/min}$$
$$T_1 = \frac{9.55 P_1}{n_1} = 9.55 \times 3.49 \times 10^3 / 384 = 86.80 \text{ N·m}$$

Shaft II:
$$P_2 = P_1 \eta_{12} = P_1 \eta_{gear} \eta_{bearing} = 3.49 \times 0.97 \times 0.99 = 3.35 \text{ kW}$$
$$n_2 = \frac{n_1}{i_{12}} = \frac{384}{4.038} = 95.1 \text{ r/min}$$
$$T_2 = \frac{9.55 P_2}{n_2} = 9.55 \times 3.35 \times 10^3 / 95.1 = 336.4 \text{ N·m}$$

Shaft III:
$$P_3 = P_2 \eta_{23} = P_2 \eta_{gear} \eta_{bearing} = 3.35 \times 0.97 \times 0.99 = 3.22 \text{ kW}$$
$$n_3 = \frac{n_2}{i_{23}} = \frac{95.1}{2.989} = 31.8 \text{ r/min}$$
$$T_2 = \frac{9.55 P_3}{n_3} = 9.55 \times 3.22 \times 10^3 / 31.8 = 967 \text{ N·m}$$

Shaft IV:
$$P_4 = P_3 \eta_{34} = P_3 \eta_{coupling} \eta_{bearing} = 3.22 \times 0.99 \times 0.99 = 3.16 \text{ kW}$$
$$n_4 = n_3 = 31.8 \text{ r/min}$$
$$T_4 = \frac{9.55 P_4}{n_4} = 9.55 \times 3.16 \times 10^3 / 31.8 = 949 \text{ N·m}$$

The above are summarized in Table 2-3.

Chapter 2 Outline Design of Transmission Equipments

Table 2 – 3 Kinetic and dynamic parameter of each shaft

Shaft No.	Power /kW	Rotating speed / (r · min^{-1})	Torque / (N · m)	Drive type	Speed ratio	Efficiency
0	3.64	960	36.21	Belt	2.5	0.96
I	3.49	384	86.80	Gear	4.038	0.96
II	3.35	95.1	336.49	Gear	2.989	0.96
III	3.22	31.8	967.80	Coupling	1.0	0.98
IV	3.16	31.8	949.00			

Chapter 3 Design Calculations for Transmission Components

Design calculation for transmission components includes the determination of materials, heat treatments, parameters, dimensions and main structures. It can provide preparation for the design of assembly sketch.

The kinetic and dynamic parameters of the transmission devices and the working conditions given by the design program are the original data for the design calculation of transmission components.

The requirements of transmission component design and the problems that should be emphasized are introduced briefly in the following.

3.1 Design Calculation for Transmission Components out of the Reducer

In the process of designing transmission devices, if there are other transmission components out of the reducer, usually these should be designed first, such as belt drives, chain drives, open gearing and so on. After the parameters of above transmission components have been decided, such as belt length L, tooth number of the sprocket, tooth number of open gear and so on, the actual transmission ratio of external drive can be determined. Then we will modify the transmission ratio of the reducer and design the reducer's internal transmission components, which can reduce the cumulative error of transmission ratios in all the transmissions.

Generally, for transmission components out of the reducer, it is only required to determine the main parameters and dimensions, rather than the details.

3.1.1 V-Belt

To design a V-belt drive, these information, such as the type, the base diameters and widths of pulleys, belt length, the center distance and so on, should be determined.

After the dimensions of pulleys are determined, it should be checked whether the drive

Chapter 3   Design Calculations for Transmission Components

size is suitable for transmission. For example, whether the external radius of the small pulley directly mounted on the motor is less than the center height of the motor; whether the radius of hub pore is equal to the radius of the motor shaft. If it is inappropriate, modify the base diameters d_{d1} and d_{d2}, and redesign them. After determine them, check the actual speed ratio of belt.

According to the installation of pulleys, design the diameter of hub pore. Generally, it should meet the standard requirements, as shown in Table 10 – 2.

3.1.2   Chain

To design a chain drive, these information, such as chain pitch, the tooth numbers of the sprockets, the diameters of sprockets, the widths of hubs, the center distance and the force acting on the shafts should be determined.

It is better that the tooth number of the sprocket is odd. In order to control outside size, the tooth number of low-speed sprocket should not be too big. The single strand should be replaced by double strands when the chain pitch is too big. In order to avoid offset links, it is better to take an even number as the link number (see Fig. 3 – 1).

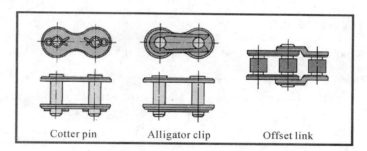

Fig. 3 – 1   Connection of chain links

3.1.3   Open gearing

To design an open gearing drive, the parameters, such as module, tooth numbers, pitch diameters, addendum diameters, tooth widths, hub lengths and forces acting on the shafts should be determined.

3.1.4   Selection of couplings

According to the calculated torque T, the permitted maximum torque of the selected coupling should be larger than the calculated torque T and the bore diameter should accord with the connected shaft diameter. The type of coupling should be determined according to working requirements.

3.2 Design Calculation for Transmission Components in the Reducer (Gear Design)

Design calculation for gear drives can refer to textbook or design book, and the following points should be paid attention to.

3.2.1 Selection of material and heat treatment

The manufacturing method of gear blank should be considered to select material for the designed gear. If the addendum diameter $d_a \leqslant 400 - 600$ mm, forged blank should be adopted. If $d_a > 400 - 600$ mm or it is difficult to forge, cast steel or cast iron should be applied.

Heat treatment can improve material properties, especially hardness, thereby raise carrying capacity. Steel gears can be classified into two kinds by the case hardness of gear tooth, those are soft tooth surface (case hardness $\leqslant$ 350HBS) and hard tooth surface (case hardness $>$ 350HBS). The harder the case hardness is, the smaller the volume is.

3.2.2 Structure of gear

Small steel gears, typically $e < 2m_t$ as shown in Fig. 3-2(a) and (b), cylindrical gear are machined directly into the surface of the shaft that carries the gear. This type of gear style is called gear shaft. Referring to Table 17-1, the structure and sizes of forged gears can be designed.

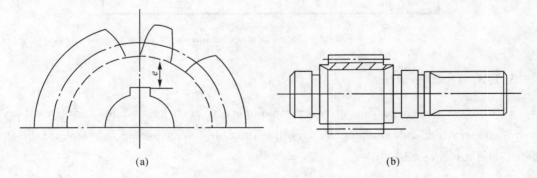

Fig. 3-2 Gear shaft

3.2.3 Center distance of gear drive

If the reducer is mass produced, the parameters such as center distance can be decided referring to standard reducers to improve interchangeability, and vice verse. To manufacture and install conveniently, it is better to take value from R40 series of Table

10-2. The center distance can be adjusted to accord to the value of R40 series by altering module or tooth number or by modifying gear for spur gear, while for helical gear, it is available to change helix angle to adjust the center distance.

3.2.4 Parameters of gear

In order to ensure the accuracy of calculation and manufacture, the helix angle of helical gear must calculate to "second" precisely, and reference diameter must reach the third place after decimal point, where arbitrary rounding is prohibited.

Chapter 4　Construction of Reducer

A reducer consists of transmission components (gears), bearings, housings and other associated components. Fig. 4 - 1 shows a two-level cylindrical gear reducer. The structure of the reducer will be introduced according to Fig. 4 - 1.

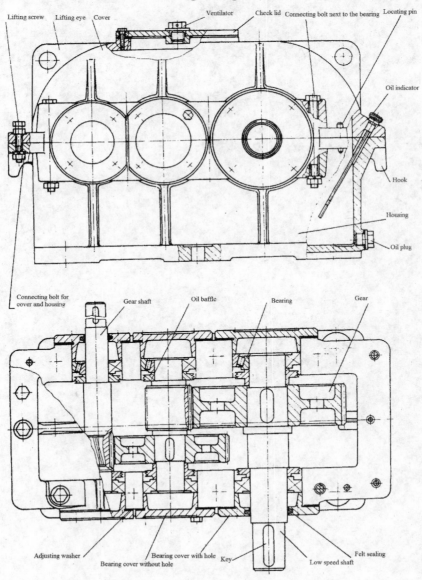

Fig. 4 - 1　Two-level cylindrical gear reducer

Chapter 4   Construction of Reducer

4.1   Gear, Shaft and Bearing Combination

Fig. 4 – 1 shows a gear shaft (the smallest gear), whose diameter approaches to the shaft diameter. The larger gear is installed on shaft radially fixed by a straight key. The components on shafts are axially fixed by shoulders, sleeves (oil blocking rings) and bearing covers (end shields). The bearings are lubricated by grease. In order to prevent the oil in the housing from entering bearings, oil blocking rings (oil baffle) should be designed in the bearing seat bore near the inside wall of housing between bearings and gears. In order to avoid the lubricant leakage at the connection of bearing cover and the extended segment of shaft and prevent the outside dirt and particles from entering the housing, sealing components should be designed in the bearing cover with through hole.

4.2   Housing

It is the base of transmission components and should have sufficient strength and stiffness.

Usually housing is made of gray cast iron, and under heavy loads with impacts, cast steel housing can be used. For single-piece, in order to simplify the process and reduce costs, welded steel housing can be produced.

Fig. 4 – 1 shows a housing made of gray cast iron. The housing is horizontally split along the axis line, so that the installation and disassembly of the components and parts are convenient. The upper cover and the lower housing are connected by general bolts. The connection bolts besides the bearing should be as close as possible to the bearing seat bore and the projecting flat near bearing should have sufficient supporting area, so that the connection bolts could be designed here and there is enough space required by wrench when the bolts are screwed tightly. In order to ensure the housing has sufficient stiffness, supporting rib should be added near the bearing seat. And in order to guarantee the stability of reducer installed on the base, the machined area of the bottom of the housing should be reduced as possible as you can, and the bottom planes of the housing are generally not a complete plane (Fig. 4 – 2 (b) is better). Fig. 4 – 1 shows a housing with 3 rectangles machined bottom planes.

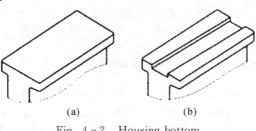

Fig. 4 – 2   Housing bottom
(a) Wrong;   (b) Correct

4.3 Associated Components

In order to ensure the normal operating of reducer, not only the structure design of gear, shaft and housing should be paid attention to, but also the oil injecting and ejecting, oil depth checking, precise location of the parts of housing during repairing and disassembling, lifting, transporting and other associated components or parts should be considered.

4.3.1 Checking hole and lid (see Table 16-1)

In order to check the engaging status, contact pitting and gear backlash and inject lubricating oil into the housing, the checking hole should be designed where the gear engaging position can be observed directly. And our hand can put into the housing through the hole to check the gear engaging performance. In Fig. 4-1, the checking hole is a rectangle on the top of the upper cover. During working the lid of the checking hole is fixed on the cover with screws. Projection should be designed on the surface of the cover contacting with the lid, in order to machine the surface easily, and a washer should be used as sealing (shown in Fig. 4-3). The lid can be made of steel or cast iron (in Fig. 4-4). In Fig. 4-4(a), it is a steel lid, the structure is simple and machining is not required. In Fig. 4-4(b), it is a cast-iron lid and many positions need to be machined, so it is not used widely.

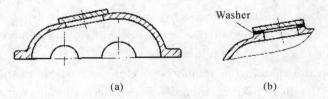

Fig. 4-3 Checking hole and lid
(a) Wrong;   (b) Correct

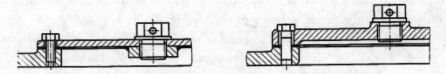

Fig. 4-4 Checking hole lids made of different materials
(a) Made of steel;   (b) Made of cast iron

4.3.2 Ventilator (see Table 16-2 and Table 16-3)

When the reducer is working, the temperature in the housing rises, and the air is expanded and pressure increases. So in order to ensure the expanded air can flow from the housing freely and avoid oil leakage, a ventilator should be designed on the top. Fig. 4-1 shows a ventilator screwing into the thread hole of the checking hole lid. Fig. 4-5(a)

Chapter 4   Construction of Reducer

shows a simple ventilator, and Fig. 4-5(b) shows a ventilator with filter screen.

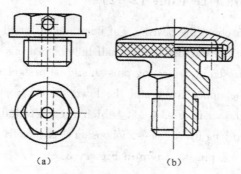

Fig. 4-5   Ventilators

4.3.3   Bearing cover and sealing (see Table 16-4 and Table 16-5)

In order to fix the axial location of the whole shaft and carry axial load, plate covers (bearing covers) should be used at the two ends of the bearing seat bores. There are 2 types of plate covers: flange type (in Table 16-4) and imbedded type (in Table 16-5). Flange type is suggested and is fixed on the housing with 4 or 6 screws. The plate cover with through hole should be used to the extended shaft end and sealing component should be installed in the plate cover.

4.3.4   Oil blocking ring (oil retainer or oil baffle) (see Table 16-7)

The functions and structures of oil blocking rings used with bearings are different according to whether grease or oil is used as lubricant (see Fig. 4-6 and Fig. 4-7). When oil is used, the ring should be installed on the high-speed gear shaft in case that the hot oil injected from gear backlash enters bearings, influencing the bearing life. When the diameter of dedendum circle is larger than bearing seat bore diameter, the ring does not need to be used.

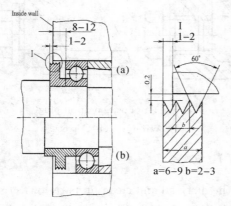

Fig. 4-6   Position of oil retainer and bearing
for grease-lubricant
(a) Correct;   (b) Wrong

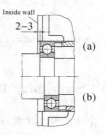

Fig. 4-7   Position of bearing
for oil-lubricant
(a) Correct;   (b) Wrong

4.3.5 Locating pin (see Table 12-2)

In order to precisely machine the bearing seat bore and maintain the original position accuracy of the top and bottom half hole of bearing seat after each assembly and disassembly, locating pins should be designed on the connection flange of the upper cover and the lower housing. Two locating tapered pins are shown in Fig. 4-1, which are installed on connection flange on the two lengthways sides of housing dissymmetrically to enforce locating effect. A pin is shown in Fig. 4-8.

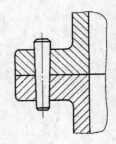

Fig. 4-8 Locating pin

4.3.6 Separating screws (see Table 11-7)

In order to enhance sealing, something such as sealing glue would be lain on the split housing planes during assembly, and the separation is somewhat difficult. So at proper position of connection flange, 1 or 2 thread holes should be processed in order that separating screws with cylindrical or flat ends can screw into them. While screwing the separating screw, the upper cover can be pushed up (see Fig. 4-9).

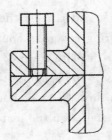

4.3.7 Oil surface indicator (see Table 15-2)

Fig. 4-9 Separating screw

In order to check the oil depth in reducer and ensure there is sufficient oil, an oil surface indicator should be designed (see Fig. 4-10). Fig. 4-1 shows an oil dipstick.

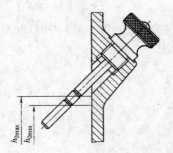

Fig. 4-10 Oil indicator

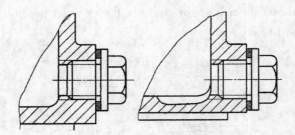

Fig. 4-11 Position of oil plug
(a) Wrong; (b) Correct

4.3.8 Screw plug (see Table 16-6)

While changing the oil, in order to eject the dirty oil and clean preparation, ejecting oil hole should be designed at the bottom of the housing below the oil bath, and it is usually plugged by a fine thread screw plug. An anti-leakage washer should be used between the plug and the housing (see Fig. 4-11).

4.3.9 Oil cup

When the rolling bearings are lubricated by grease, the grease should be replenished regularly. To complement the grease conveniently, oil cup is refueled in the bearing seat of housing, used for injecting the grease (see Fig. 4 – 12).

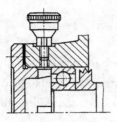 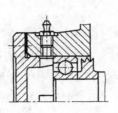 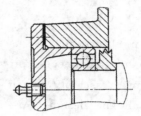

Fig. 4 – 12 Oil cup

4.3.10 Lifting equipment (see Table 16 – 8)

If the reducer weight exceeds 25 kg, lifting equipment should be designed on the housing for convenient transport, such as lifting eye or lifting hook. In Fig. 4 – 1, there are 2 lifting eyes on the upper cover and 4 lifting hooks on the lower housing.

Chapter 5   Assembly Sketch Design

Assembly drawing is used to convey the relationship, structural shapes and dimensions of components, providing basis for assembly, debugging and maintenance of the machine. Assembly sketch is not only the usual beginning, but also the important stage of all the design. At this stage, strength, stiffness, manufacturing process, assembly, debugging and lubrication of each part should be taken into integrated account, because the structure and dimensions of most components are determined here.

The initial purpose of sketch design is to check whether the tentative kinetic parameters and the structures and dimensions of transmission components cooperate or interference. At the same time, the structures, spans and force action points of the shafts can be determined to check the shaft strengths and rolling-contact bearing lives. And the final purpose is to determine the structures and dimensions of all the components and parts, providing basis for working drawing design of assembly and components.

During this stage, we inevitable need to modify our design again and again, in order to obtain a better structure. Reducer assembly sketch design can be divided into two stages, i. e. tentative drawing and completed drawing.

5.1   Tentative Drawing of Assembly Sketch

5.1.1   Preparation

To comprehend the function and structure of each component and part deeply, please read and understand a typical reducer assembly drawing, watch some video about reducer, and disassemble or assemble a physical reducer.

1. Determine the main dimensions of gear drive

Including center distance, reference circle and addendum circle diameters, gear widths, hub length and so on.

2. Look up the installation dimension of selected motor

Such as motor shaft extending diameter D, length E and the center height H.

3. Select coupling type

According to required function of transmission system, proper coupling type can be selected.

Chapter 5 Assembly Sketch Design

If the motor and the reducer are installed on a public base, concentricity between the two shafts can be ensured easily, so it is inessential to choose coupling with high compensation ability. Moreover, to reduce starting load, the coupling connecting with high-speed shaft should possess a small moment of inertia and good damping property. Table 14-1 to Table 14-3 show some elastic couplings.

For the coupling connecting the reducer and the working part, because it is on low-speed shaft, it is not required to have a small moment of inertia. If the reducer and the working part are installed on the same base, elastic coupling can be adopted frequently (as shown in Table 14-1 to Table 14-3). It they are not, this coupling should have a higher compensation ability, so maybe cross-sliding coupling should be used.

4. Select rolling-contact bearing types

Determining the initial type of bearing based on properties and magnitude of load, rotating speed and working requirements. For spur gear, deep groove ball bearings are recommended firstly. If the bearing seat endures a radial loads (R) and a big thrust loads ($A > 0.25R$), it is better to use angular contact bearings or tapered roller bearings which are the most popular ones. Refer to Table 13-1 to Table 13-3 to select the types.

5. Determine the lubrication and sealing of rolling-contact bearings

When the peripheral velocity of the transmission component dipping into the bath $v \leqslant 2$ m/s, grease should be used as lubricant. When $v > 2-3$ m/s, the oil splashed by the rotating transmission component should be used as lubricant and oil groove should be designed (see Fig. 5-1). Fig. 5-2 shows a bearing cover with oil channel.

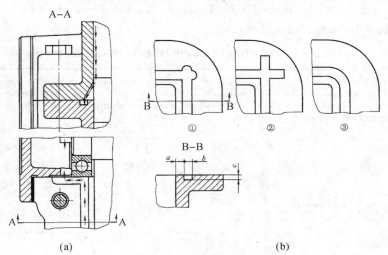

Fig. 5-1 Oil path and oil groove for oil-lubricant
(a) Oil path; (b) Structure and dimensions of oil groove
①By cylindrical milling tool; ②By plate milling tool; ③By casting
$b = 6-8$ mm; $c = 3-5$ mm; $a = 4-6$ mm (machined), $a = 5-8$ mm (casting)

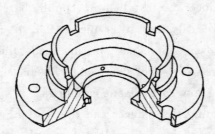

Fig. 5-2 Bearing cover with oil channel

Design the sealing type of the bearing cover according to lubrication and working conditions.

6. Determine the structure and dimensions of housing

Reducer housing is the base to support and install the transmission components, such as gears, so good rigidity is necessary to avoid too much deformation and nonuniform load distribution on the gear. To reinforce the rigidity, ribs are arranged on the flange of bearing seat.

If the reducer is mass produced, housing is generally made of cast iron (such as HT150, HT200). If it is single, welded steel plates can be employed to reduce the mass and overall dimensions.

The dimensions of housing should be determined by experiment formulas, then the values should be rounded off and maybe some need to be modified according to structure demand. For the dimensions related to standard components, corresponding standard values should be taken (diameters of bolts, screws and pins). The dimensions should be determined according to the formulas in Table 5-1. Fig 5-3 shows a typical casting housing structure (of our course design).

Table 5-1 Structural and dimensions of cast housing

Name	Symbol	Recommended relationship of dimensions	
Wall thickness of lower housing	δ	Two-level gear reducer: $\delta=0.025a^*+3 \geqslant 8$	Worm gear reducer: $\delta=0.04a^*+3 \geqslant 8$
Wall thickness of cover	δ_1	$\delta_1=0.9\delta \geqslant 8$	Worm below: $\delta_1=0.85\delta \geqslant 8$ Worm above: $\delta_1=\delta \geqslant 8$
Flange thickness of split surface of lower housing	b	$b=1.5\delta$	
Flange thickness of split surface of cover	b_1	$b_1=1.5\delta_1$	
Foot thickness of anchor bolt	p	$p=2.5\delta$	
Thickness of rib on lower housing	m	$m>0.85\delta$	
Thickness of rib on cover	m_1	$m_1>0.85\delta_1$	

Chapter 5 Assembly Sketch Design

Continued

Name	Symbol	Recommended relationship of dimensions		
Sum of two center distances of two-level cylindrical gear reducer	a_1+a_2	$\leqslant 300$	$\leqslant 400$	$\leqslant 500$
Diameter of bolt (or screw) next to the bearing	d_1	M12	M16	M20
Diameter of bolt's through hole next to the bearing	d'_1	13.5	17.5	22
Diameter of bolt's countersunk seat next to the bearing	D_0	26	32	40
Flange dimension of boss next to the bearing (wrench space)	c_1	20	24	28
	c_2	16	20	24
Diameter of bolt (or screw) connecting lower housing with cover	d_2	M10	M12	M16
Diameter of bolt's through hole connecting lower housing with cover	d'_2	11	13.5	17.5
Diameter of bolt's countersunk seat connecting housing with cover	D_3	24	26	32
Flange dimension of housing (wrench space)	c_1	18	20	24
	c_2	14	16	20
Diameter of anchor bolt	d_φ	M16	M20	M24
Diameter of anchor bolt's hole	d'_φ	20	25	30
Diameter of anchor bolt's countersunk seat	D_φ	45	48	60
Flange dimension of foot (wrench space)	L_1	27	32	38
	L_2	25	30	35
Number of anchor bolts	n	Tow-level gear	6	
		Worm gear	4	
Screw diameter of bearing cover	d_3	Shown in Table 16-4		
Screw diameter of checkinghole cover	d_4	M6		M8
Diameter of tapered locating pin	d_5	10	12	16

Continued

Name	Symbol	Recommended relationship of dimensions
Height of reducer center	H	$H \approx (1 - 1.12)a^*$
Height of boss next to the bearing	h	Determined by structure according to D_2 and c_1
Radius of boss next to the bearing	R_δ	$R_\delta \approx c_2$
Outer diameter of bearing cover	D_2	Shown in Table 16-4
Distance of bolts next to the bearing	S	$S = D_2$
Distance from outer wall of housing to the end of bearing seat	K	$K = c_1 + c_2 + (5 - 8)$
Distance from inside wall of housing to the end of bearing seat		$K + \delta$
Distance between addendum circle of big gear with inside wall of housing	Δ_1	$\Delta_1 \geqslant 1.2\delta$
Distance between the end of gear with insider wall of housing	Δ_2	$\Delta_2 \geqslant \delta$

Note: a^* — center distance of low-speed level.

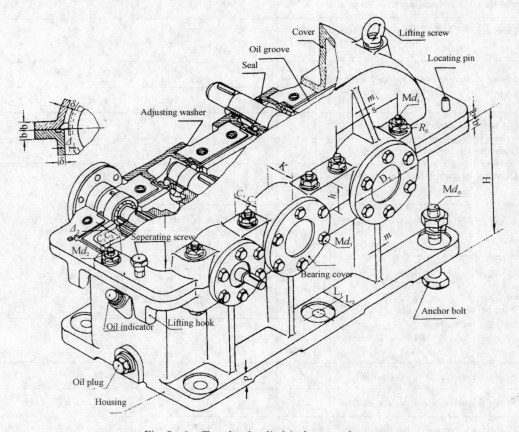

Fig. 5 - 3 Two-level cylindrical gear reducer

5.1.2 Procedure of Sketch Drawing

The procedure is introduced by instancing a two-level cylindrical gear reducer, as shown in Fig. 5-4.

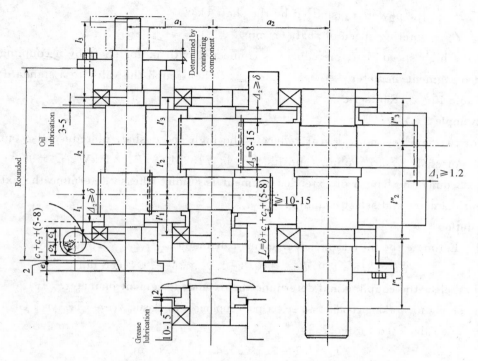

Fig. 5-4 Sketch of a two-level gear reducer

(1) Select the scale. Generally, scale of 1:1 and A0 paper should be adopted and used. Three views are popularly used to convey an assembly drawing and the centerlines should be drawn on the paper first.

(2) On the top view, draw the outline dimensions of gears such as the addendum circle and gear width. To ensure the gears mesh fully, the smaller gear width should be larger than the bigger one by 5-10 mm, that is $b_1 = b_2 + (5-10)$ mm.

The distance between two level transmission components $\Delta_3 = 8 - 15$ mm.

(3) Draw the inside wall lines of housing. The distance between the pinion end surface and the inside wall line $\Delta_2 \geqslant \delta$, so two inside wall lines along the length direction of housing are determined. Then according to $\Delta_1 \geqslant 1.2\delta$, we can obtain the inside wall line close to the low speed level larger gear along the width direction of housing and that close to high-speed-level smaller gear can be determined by front view.

(4) Tentative determination of shaft diameters.

1) Determine the extended segment diameter of high-speed shaft.

If a belt pulley is installed on the extended segment, the diameter of shaft can be calculated by the following formula.

$$d \geqslant A_0 \sqrt[3]{\frac{P}{n}} \text{ mm} \qquad (5-1)$$

where A_0 — allowable torsional and shear stress coefficient, usually in the range of 110 – 160;
P — the power transmitted by the shaft (kW);
n — rotating speed of shaft (r/min).

If the high speed shaft of reducer is connected with motor shaft by a coupling, the extended segment diameter $d = (0.8 - 1.0)d_{motor}$, then round the value to a standard value (see Table 10 – 2, R40 series).

Example 5 – 1

For two-level cylindrical gear reducer, the high speed shaft of reducer is connected with motor shaft by a coupling, and its type is Y132M1-6, $P_0 = 4$ kW, $n_0 = 960$ r/min, $d_{motor} = 38$ mm, the length of extended segment is 80mm. Please determine the extended segment diameter and select coupling.

Solution

Ⅰ. Estimate the initial extended segment diameter.

$$d = (0.8 - 1.0)d_{motor} = (0.8 - 1.0) \times 38 = 30.4 - 38 \text{ mm}$$

Ⅱ. Select the coupling and determine the extended segment diameter.

According to working conditions, select pin coupling with elastic sleeve (TL type, GB 4323—2002). The calculated torque T_c is

$$T_c = KT = 1.5 \times 39.8 = 59.7 \text{ N} \cdot \text{m}$$

where T—nominal torque transmitted by coupling, here

$$T = 9.55 \frac{P}{n} = 9.55 \times \frac{4 \times 10^3}{960} = 39.8 \text{ N} \cdot \text{m}$$

K—work coefficient, here $K = 1.5$.

For TL6: $T_n = 250$ N · m $> T_c = 59.7$ N · m, permitted speed $[n] = 3,300$ r/min $> n_0 = 960$ r/min, $d_{min} = 32$ mm and $d_{max} = 42$ mm.

If the extended segment diameter $d = 32$ mm, then $d_1 = d_{motor} = 38$ mm, $d_2 = d = 32$ mm. So TL6 can meet the requirement.

2) The extended segment diameter of the low speed shaft is calculated by formula (5 – 1), then rounded off. If a coupling is installed on the extended segment, according to calculated torque T_c and the basic diameter, select proper coupling. If a sprocket is installed on it, the diameter of the extended segment should be the same with the hole diameter of the sprocket.

3) The basic diameter of the middle shaft is calculated by formula (5 – 1). Generally, the inner diameter of bearing on middle shaft should not be less than that of high speed shaft.

(5) Structure design of shaft.

Shafts of gear reducer are made into stepped shaft, as shown in Fig. 5 – 5. But the number of steps should be as less as possible.

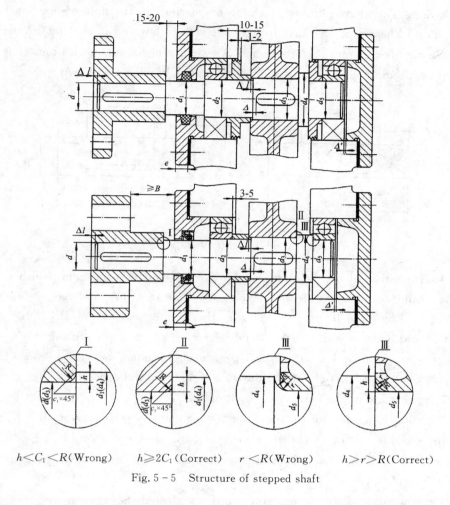

$h<C_1<R$(Wrong) $h \geqslant 2C_1$(Correct) $r<R$(Wrong) $h>r>R$(Correct)

Fig. 5−5 Structure of stepped shaft

1) Determine radial dimensions. Dimensions of locating shoulders should be a little larger, as shown in Table 10−3. Transition fillet radius is shown in Table 10−4. The dimensions of locating shoulder for bearing inner race are shown in Table 13−1 to Table 13−3. The felt sealing is a standard component and d_1 should be in Table 15−3 and Table 15−4. Diameter variation of the shoulder whose purpose is to be assembled and disassembled conveniently should be smaller.

2) Determine axial dimensions. If the shoulder is applied to fix parts and endure axial force, the changed end of shaft should be leveled with part's end. There should be a small distance ($\Delta l = 2-3$ mm) between the end of shaft segment and the end of sleeve or hub to ensure reliable positioning. The length of straight key should be less than the hub width by 5−10 mm, then round it to a standard value according to Table 12−1.

(6) Tentative selection of bearing and determination of bearing location.

According to the radial dimensions of shafts, tentative bearings can be determined and the two bearings on a shaft should be the same. Fig. 5-4 and Fig. 5-5 show the bearing sketch drawing. Then determine the location of bearings in the hole of housing based on the lubrication of bearings. Distance from inside wall of housing to the end of bearing seat is named S. S=10-15 mm for grease and S=3-5 mm for oil lubrication. The outline drawing of deep groove ball bearing and tapered roller bearing are shown in Fig. 5-6.

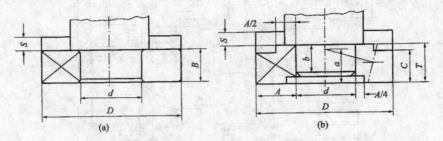

Fig. 5-6 Outline drawing of bearing

(7) Determination of the bearing seat bore width and housing edge width (see Fig. 5-4). According to diameter of bolt next to the bearing (Md_1) and flange dimension of housing required by wrench space (c_1 and c_2), the bearing seat bore width $L=\delta+c_1+c_2+(5-8)$ mm. And housing edge width should be $\delta+c_1+c_2$, as shown in Fig. 5-3.

(8) Drawing of all the bearing covers, connecting bolts and adjustable washers.

One of connecting bolts should be drawn completely, and others only by centerlines.

(9) Determination of the extended length of shaft.

If the diameter of the hub on the extended segment is relatively small, the distance between the shoulder of the extended segment and the plate cover could be 5-10 mm, as shown in Fig. 5-5. Otherwise, the distance should not be shorter than the length of the screw on the bearing cover.

(10) Determination of the force points on the shaft and the components.

From the sketch, every force point position and all the distances can be determined, such as $l_1, l_2, l_3, l'_1, l'_2, l'_3$, and l''_1, l''_2, l''_3 for the transmission components, the midpoint of the width should be taken as force point. For angular contact bearings, the force point is the pivot point (see Fig. 5-6) whose distance to thick edge of the outer race is shown in Table 13-2 and Table 13-3.

5.2 Strength Calculation of Shafts, Bearings and Keys

5.2.1 Strength calculation of shafts

The strength calculation of shafts should be done referring to the textbook. And if after calculation, the strength is not sufficient, the diameter should be increased.

5.2.2 Life calculation of rolling-contact bearing

It is better that bearings' life approximately accord with reducer's life which is general 36,000 h, and the lowest life for bearings is 10,000 h. After calculation, if the life can not meet the demand, modify the type of bearings to improve the life.

5.2.3 Strength calculation of keys

Calculate the extrusion stress and compare the calculated value with the allowable stress of the weakest component among key, shaft and hub.

5.3 Completion of Assembly Sketch Design

During this stage, the main task is to design shaft-related components and parts, housing and associated components.

5.3.1 Structure design of the shaft-related components and parts

(1) The detailed structure and dimensions of gears are shown in Table 17-1.
(2) Detailed structure of bearings (Table 13-1 to Table 13-3).
(3) Detailed structure of bearing plate covers (Table 16-4).
(4) Detailed structure of seals (Table 15-3 to Table 15-6).
(5) Detailed structure of oil blocking rings (Table 16-7).
(6) Detailed structure of sleeves and end plates (Table 11-10).

5.3.2 Structure design of housing

(1) Height of projected flat of connecting bolts near bearings h.

The distance between connecting bolts $S \approx D_2$, where D_2 is the outside diameter of plate cover (see Fig. 5-7). For the largest bearing, satisfying the size c_1, then h can be determined by drawing. Then h should be rounded to a larger integer (R20 series in Table 10-2).

(2) The distance between connecting bolts on housing edge should not be larger than 100-150 mm.

(3) The height of housing H.

$H \geqslant (d_{a2}/2) + (50-30)$ mm $+ \delta + (3-5)$ mm. d_{a2} is the addendum circle diameter of large gear.

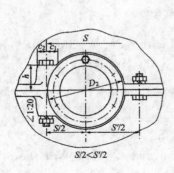 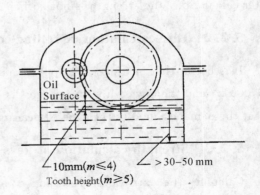

Fig. 5 – 7 Projected flat of connecting bolts near bearings

Fig. 5 – 8 Oil surface and depth of reducer

The oil depth should be determined according to Fig. 5 – 8. Then based on calculated H, the oil volume V can be calculated. V should be larger than V_0. For single level reducer, $V_0 = 0.35 - 0.7$ dm^3. For multiple level, V_0 will increase proportionally.

(4) The oil groove (see Fig. 5 – 1 and Fig. 5 – 2).

(5) The processing of housing.

1) As shown in Fig. 5 – 7, pay attention to the taper of 1 : 20 or 1 : 10.

2) At the housing surfaces contact with the bearing cover, ventilator, lifting screws and oil plug, projected flats should be designed and the height should be 5 – 8 mm. The height of projected flat for checking hole should be 3 – 5 mm. The surfaces of housing edge contacting with the connecting bolt head or nut of the upper and lower housings should be countersunk and the depth is 2 – 3 mm.

5.3.3 Associated component design

(1) Checking hole and lid. The checking hole is located on the top of transmission components. Generally the checking hole is covered by a lid and a washer is used here to prevent leakage. The lid is made of steel, iron or glass.

(2) According to oil surface indicator shown in Table 15 – 2, the inclined angle $\geqslant 45$ ℃.

(3) Ventilator, in clean environment, see Table 16 – 2; in dirty environment, see Table 16 – 3.

(4) Oil plug is shown in Table 16 – 6. Cylindrical fine thread or taper thread is designed in the oil plug. For cylindrical thread, a sealing washer should be used together.

(5) Lifting equipment is shown in Table 16 – 8.

(6) Separating screws are shown in Table 11 – 7.

Chapter 5 Assembly Sketch Design

(7) Locating pins are shown in Table 12-2, the length$>b+b_1$.

5.3.4 Determination of fit type and fit dimensions

According to chapter 18, the fit types and fit dimensions can be determined.

5.3.5 Checking

See Table 5-2 and Fig. 5-9.

Table 5-2 Correct and wrong design of housing and associated components

Wrong	Correct
Several projected flats are not machined continuously. It is not convenient for mold pulling	Several projected flats are machined continuously and the heights are the same. It is convenient for mold pulling
Casting surfaces of housing edges connecting with bolt head and washer are not machined. The bolt is likely to bear eccentric load	Countersunk is machined at the surface of housing edges connecting with hold head and washer
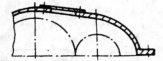 There is no projected flat at the checking hole of housing, the checking hole is too far away from the meshing zone, and there is no washer under the lid	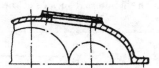 There is a boss on the housing, checking hole is just over the meshing zone, and there is a soft washer under the lid

Continued

Wrong	Correct
The inclined angle of the oil indicator hole is too big, the hole can not be machined and the indicator can not be assembled	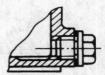 The height and inclined angle are rational
 Oil draining hole is too high	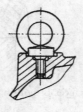 The minor diameter is a little lower than the housing bottom and there is a groove on the housing bottom, so that the oil can be drained thoroughly
There is no countersunk seat on the supporting surface, the screw can not be screwed into the housing completely; turn number of screw connection is not enough and there should be a boss on the internal surface of the housing	The boss on the external surface is higher, there is a boss on the internal surface; there is a countersunk seat on the top surface
 The machined surface is too big	 The machined surface is reduced

Chapter 5 Assembly Sketch Design

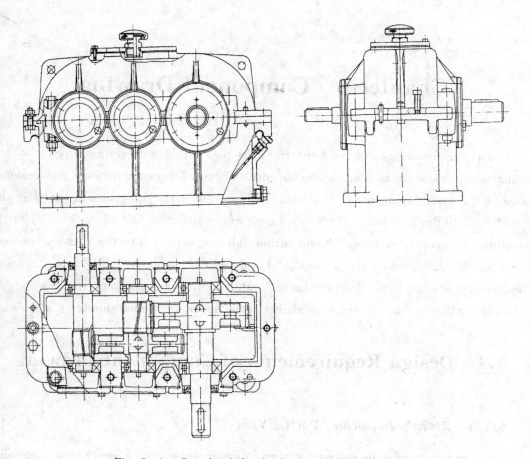

Fig. 5-9 Completed sketch of a two-level gear reducer

Chapter 6 Component Drawing

Component drawings are the basic technical files for manufacturing, inspection and technological procedure making of components. They are drawn separately and designed based on the assembly drawing. It should not only reflect the designer's intent, but also should indicate the possibility of manufacturing and installation and the rationality of the structure. Component drawing should include all the details that the manufacture and inspection need. Views, dimensions and tolerances, geometric tolerances, surface roughness, materials and heat treatment requirements, and various specifications of technical conditions that has not been shown in the above-mentioned should be included.

6.1 Design Requirements of Component Drawing

6.1.1 Basic View, Scale, Partial View

Each component must be separately drawn in a standard sheet. A set of views (including basic view, profile view, partial view and other specified representations) should be used reasonably to express the structure and dimensions completely, accurately and clearly. Scale should be 1 : 1 as much as possible, in order to enhance the reality sense of components. If necessary, it can be appropriately enlarged or reduced according to the standards. For the detail structure of components (such as relief groove, transition fillet and the core holes required to be kept, etc.), if necessary, partial enlargement can be used.

The arrangement of drawing should be made according to the outline sizes of the component. And dimensioning, technical condition writing and title bar drawing also should be concerned as well.

Basic structure and main dimensions of the component should be drawn according to the assembly drawing, which should be consistent with the assembly drawing and not be altered arbitrarily. If it has to be altered, the assembly drawing should be revised correspondingly.

6.1.2 Dimensioning of Component Drawing

Dimensions and tolerances on the drawing, as the basis of machining and testing, must be complete, accurate and reasonable. The method of dimensioning and tolerance indicating should be in accordance with the standards and the requirements of the machining process, and facilitate testing. In the dimensioning of component drawing, following points should be paid attention to.

(1) Correctly select the datum plane and line.

(2) Most of dimensions should be placed as intensively as possible in a view that can reflect the characteristics of the component's structure.

(3) The drawing should include enough dimensions and tolerances required by the processing and testing, in order to avoid any conversion in machining process.

(4) All dimensions should be placed outside the view as possible as you can. The extension lines for dimension can only link to the visible lines of component. Try not to make lines cross.

(5) For the dimensions of fit and position of high precision, it is necessary to refer to the related tolerance table, according to the fitting property and accuracy grade identified in the assembly drawing, and then note the limit deviation of each dimension. For example, in an assembly drawing, the fit dimension of a shaft and a gear hole is 60H7/n6, you can use n6 to find out, in the tolerance table, the limit deviations of shaft are $+0.039$ and $+0.020$, then mark it with $\phi 60^{+0.039}_{+0.020}$ in the component drawing of shaft. For the gear hole, you can use H7 to find out the upper and lower deviations are $+0.030$ and 0, then mark it with $\phi 60^{+0.030}_{0}$ in the component drawing of gear. Besides, diameter of the extended segment, center distances on the housing, dimensions of keyways on a shaft, and so on, are required to be written on the drawing with the corresponding limit deviations.

6.1.3 Indication of Surface Roughness

Whether surface roughness is selected appropriately or not, will influence abrasion resistance, corrosion resistance, anti-fatigue ability and fitting properties of a component's surface, as well as the processing technique and the manufacturing cost. Thus, before determining the surface roughness, consider it carefully, according to the work requirements of components, precision grade and processing method. Under the condition that the normal work is not affected, make R_a as larger as possible to reduce the manufacturing cost.

Usually analogy method is used to determine height parameter value of component's roughness. Refer to the recommended values in Table 18-8.

All surface roughness of components must be shown. If many surfaces have the same roughness, roughness can be written once in the top right corner of the component drawing, with the word "others", which can make the drawing more clear and tidy.

6.1.4 Indication of Shape and Location Tolerance

As important criteria of evaluating part's processing quality, necessary shape and location tolerances should be written on the drawing.

Because the performance requirements of different parts differ, the kinds and grades of shape and location tolerances are not same.

6.1.5 Technical Conditions

The technical requirements which must be guaranteed are inconvenient to be expressed by graphics or symbol, can be written briefly in the technical conditions. Technical conditions of different parts are different, and can be edited according to the following:

(1)Requirements for casting and forging blank. For example, some requires no burr or oxide coating on the forging blank's surface; some requires aging treatment of housing before machining and so on.

(2)Requirements for aging treatment method and hardness.

(3)Requirements for machining. For example, whether it is needed to keep center bore, location pin hole of housing generally requires boring and reaming after the matching of upper and lower housing.

(4)Other requirements. For example, statement about the chamfer and fillet without dimensioning, requirements for modification, such as painting and chrome plate, static and dynamic balance test for large rotary part of high speed.

(5)Title block. Place title block at the bottom right corner, in order to state name of the component, material, quantity, figure number, scale and the author's name.

6.2 Shaft Drawing Design

6.2.1 View

Shaft is a combination of coaxial cylinders and cones, with keyway, relief groove, collar, shoulder, thread and center hole. So just one view, making the axis horizontal and the keyway top, can fully express the basic feature of a shaft. Then draw partial profiles of detailed features, such as keyway, relief groove, hole, grinding undercut. If necessary, enlarge partial profiles.

6.2.2 Dimensioning

In order to ensure the axial location of components on the shaft, before indicate each segment's length, determine the main datum and auxiliary datum according to the requirement of design and process, and select the reasonable form of indication. To avoid creating closed dimension chain, the length of higher precision segment should be indicated, while the lengths of lower ones should be abandoned.

6.2.3 Indication of tolerance and surface roughness

Tolerances of important dimensions, such as diameters of the segments to be mounted with gear, sprocket, coupling and bearing, should be indicated according to the fitting property and accuracy grade identified in the assembly drawing. Dimensions and tolerances of keyways should also be indicated according to the rules about key joint.

6.2.4 Technical conditions

Refer to 6.1.5. A shaft drawing is shown in Fig. 20 – 1.

6.3 Gear Drawing Design

6.3.1 Views

As a rotary component, gear can be expressed by one or two views. In the front view, with the axis horizontal, full section or half section can be used to show the structures of hole, keyway, wheel hub, spoke and rim; draw entire side view to show the shape of gear and the structures of hole, keyway, wheel hub and rim, or just draw partial view to show the dimensions of hole and keyway. All in all, it is similar to shaft drawing design.

6.3.2 Indication of dimension, tolerance and surface roughness

Select the axis as the baseline, end face as dimensional datum of the tooth width direction. Attention should not only be paid to avoiding omissions, such as the fillet, chamfer, slope, taper and keyway size, but also to avoiding duplications.

Pitch circle diameter is a basic size for design and calculation, such sizes as addendum circle diameter, hub diameter, spokes (or web), are indispensable to the processing. All of them should be indicated in the drawing. Dedendum circle diameter, as the result of the processing based on other parameters, should not be indicated.

All fit dimensions or dimensions of higher precision requirement should be indicated with tolerance and surface roughness.

As an important datum of processing, inspection and assembly, the shaft hole of gear has a higher precision requirement for the diameter. According to the fitting property and accuracy grade identified in the assembly drawing, it's necessary to look up the value in the tolerance table and indicate the limit deviation value.

The shape and location tolerances need to be indicated for gears, and the symmetrical degree tolerance of keyway's side faces related to the center line should be selected according to precision grades 7 – 9.

Besides, surface roughness should be indicated on all surfaces of gear.

6.3.3 Table of engagement property

Placed at the upper right corner, table of engagement property includes: the basic parameters of gears (modulus m, number of teeth z, pressure angle α, helical angle β), accuracy grades and tolerances of inspection items. About the accuracy grade, 7 or 8 grade can be selected.

Fig. 20 – 2 is an example of transmission components (gear).

6.4 Housing Drawing Design

6.4.1 Views arrangement

To clearly express different parts of the structure, usually more than three views are used. If necessary, sections, direction view and local enlarged view should be added.

6.4.2 Dimensioning

(1) To facilitate processing and measurement, it's better to select processing datum as dimensional datum. For example, dimensions along the height direction of the housing and cover should be based on subdivision surface (processing datum), dimensions along the width direction of the housing based on the symcenter line of width, dimensions along the length direction of the housing based on the center line of the bearing seat hole.

(2) Dimensions of housing can be divided into shape dimensions and location dimensions. Shape dimensions, including wall thickness, fillet radius, height, width and depth of the housing, groove's depth and width, all kinds of hole diameter and depth, the size of the threaded hole, etc., should be dimensioned directly. Location dimensions, such as the distance from center line of hole, central position of curve, or plane of other relevant part to the datum, should be started from the datum (or auxiliary datum).

(3) Dimensions related to working performance (such as center distance and the deviation) should be dimensioned.

(4) Deviation of fit dimensions should be dimensioned.

(5) All dimensions of fillet, chamfer, draft angle should be dimensioned or indicated in technical condition.

(6) After location dimensions of every basic part being dimensioned, other dimensions should be dimensioned from their datum.

(7) Avoid creating closed dimension chain.

6.4.3 The shape and location tolerances and surface roughness

See Table 18 - 7 and Table 18 - 8.

6.4.4 Technical condition

Refer to 6.1.5.

Chapter 7 Assembly Drawing Design

Assembly drawing is not only the guiding technical documents of the manufacture, installation, use and maintenance, but also the important material for technical personnel to understand and study the principle of the mechanical structure. Thus, assembly drawing should show the possibility, order, adjustment, operation and maintenance methods of mechanical assembly and disassembly of each component clearly and accurately.

After the assembly sketch design, the structure and assembly relation of reducer components had been determined. Then in the following component drawings, there may be some modifications, so during the assembly drawing design, we should analyze and modify them again.

7.1 Views

In assembly drawing two or three views should be taken as primary, and necessary section views or partial views as auxiliary. Try to express the working principle of gear reducer and the main assembly relationship in a basic view as far as possible. Generally, the top view of gear reducer without cover should be taken as the basic view. For worm gear reducer, main view should be taken as the basic view.

Views should be intact and clear, avoiding dotted line to show the shapes of components. If internal or detail structure is necessary, the partial section view or direction view can be used.

In the section view, to distinguish the adjacent parts, direction or distance of their section lines should be different. For the components, whose section width is smaller (<2 mm), the section line can be replaced by blackening. The direction and distance of the section lines on the same components in each view must be consistent.

Some structures can be drawn briefly according to the drawing standard.

7.2 Dimensioning

As the reference of reducer assembly, assembly drawing should have the location dimensions of related components, the overall dimensions of reducer, fit relationship and fit dimension between components, etc. Shape dimensions and tolerances of components are dimensioned only on component drawing.

7.2.1 Characteristic dimension

The dimensions showing the performance, specifications and features of the reducer should be added, such as the center distance and its deviation of transmission components.

7.2.2 Fit dimension

Refer to 5.3.

7.2.3 Installation dimension

Reducer needs not only to be installed on the foundation or mechanical equipment, but also to be connected with motor or other transmission parts, which requires some installation dimensions, such as the dimensions of its ground surface, the diameter and center distance of anchor bolt, the location dimension of anchor bolt hole, the fit length and diameter of driving shaft and driven shaft overhanging, the center height of overhanging, the distance between shaft overhanging end face and the datum of housing, etc.

7.2.4 Overall dimension

Overall dimensions show the space reducer occupies, such as overall length, overall width and overall height taken as the references of package, transportation and installation.

Layout of dimension lines should be tidy and clear, and dimensions should be concentrated to the views showing the main structure relations as many as possible; most dimensions should be placed outside the reducer, and digital should be written neatly.

7.3 Component No., Title Block and Component List

To facilitate drawing reading, assembly and production arrangement, it is necessary to number each different part and/or component and edit the title block (title bar) and component list.

7.3.1 Component No.

Numbering should conform to the national mechanical drawing standards, and avoid omissions and repetitions. Components should be arranged in sequence. Each component with the same shape, size and material should have the same number. Leader lines should extend outside the view with thin lines, should not intersect or parallel, especially should not be parallel to the profile line. For component group with obvious assembly relationship, such as bolts, nuts and washers, a common leader line can be used, but they should be numbered respectively.

Although each independent part, such as rolling bearing, ventilator and oil indicator, is composed of several components, it should have only one number. Numbers should be arranged outside the reducer, along the horizontal direction and vertical direction, with a clockwise or counter-clockwise order. Word should be neatly, and its character should be one or two size higher than the dimension's.

7.3.2 Title block and component list

Title block should be placed at the lower right corner, showing name, scale, quantity, weight, and drawing code.

Component list should indicate the No., name, quantity, material, specification, etc. of each components and parts. The process of filling in the component list is also the final process of determining the components and parts and their materials. Try to reduce the variety of materials and the species of standard components.

Component list should be filled in from bottom to top. The standard components should be written according to the rules. Material brands should be indicated.

The formats and filling methods of title block and component list are shown in Fig. 20-3 to Fig. 20-5.

7.4 Technical Feature of Reducer

Technical feature of reducer usually be placed at the blank of the drawing by a table such as Table 7-1.

Table 7-1 Technical feature of reducer

Input power/ kW	Input speed/(r·min^{-1})	Efficiency/ (%)	Total ratio i	Transmission feature						
				Level	m_n	z_1	z_2	β	Accuracy	
				High speed					Pinion	
									Larger gear	
				High speed					Pinion	
									Larger gear	

Chapter 7 Assembly Drawing Design

7.5 Technical Condition

Technical condition is determined based on the design requirement. It should be written in appropriate place with the content about assembly, adjustment, maintenance and inspection.

7.5.1 Requirements for the contact spots of gear and worm drive

Contact spot is determined by the accuracy grade of transmission component. Refer to Table 18-9. To inspect it, paint the active tooth surface and then turn it. After 2-3 laps, from the coloring of the driven tooth surface, analyze the position and the size of the contact area, and check whether it meets the precision requirement.

If not, adjust the meshing position, or appropriately scrape and grind the tooth surface and conduct load running-in, to improve the assembly precision.

7.5.2 Requirements for installation and adjustment of rolling bearing

A certain clearance must be guaranteed during the working process of rolling bearing. Too big clearance may lead to the movement of shaft. If the clearance is too small, bearing running resistance will increase, affecting its normal work, even the bearing will be stuck and damaged. For the bearing whose clearance is nonadjustable (deep groove ball bearing), leave an appropriate gap ($=0.2-0.5$ mm) between end face of outer ring and bearing cover. The greater the span is, the bigger the clearance should be. If the bearing's clearance is adjustable (angular contact ball bearings and tapered roller bearings), the clearance value should be smaller. The axial clearance adjustment method is sketched as follows:

As shown in Fig. 7-1, after the bearing combination being put into the housing bearing hole, install bearing cover on the bearing seat without adjusting shim, and then tighten the connecting screw of the cover to make the cover face press against the bearing outer ring end face. When you feel difficult to rotate the shaft, which means the internal clearance is completely eliminated, use the feeler gauge to measure the clearance δ (see Fig. 7-1(a)) between bearing cover flange and the end of the housing bearing seat. Then remove the cover, put adjusting shim (thickness $=\delta+\Delta$) between the bearing cover and the housing, finally remount the cover and screw down the screws (see Fig. 7-1(b)).

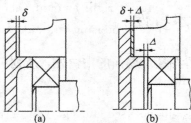

Fig. 7-1 Clearance adjustments for rolling-contact bearing

7.5.3 Requirements for gear backlash

Requirements for gear backlash should be regulated by using the maximum backlash j_{nmin} (or j_{tmax}) and the minimum backlash j_{nmax} (or j_{tmin}).

In the check of backlash, feeler or method of pressing lead wires (put the lead wire on the alveolus, then turn the gear to flatten the lead wire, sum of the thicknesses of squished lead wires on both sides of tooth is the backlash size) is used.

If the backlash does not conform to the requirements, adjust the position of transmission. For bevel gear drive, the position of bevel gear can be adjusted by changing the thickness of the gaskets to make the two cone vertexes coincide. For worm gear drive, adjust the gaskets between housing and worm gear bearing cover (add at one end and subtract at the other end), making the middle plane (main plane) of worm gear pass through the axis of the worm.

7.5.4 Requirements for the joining surfaces of cover and housing

Gasket is not allowed to be used here. If necessary, smear sealing glue or sodium silicate can be used. Before tightening the connecting bolt, use feeler (0.05 mm) to check its seal. Oil leakage is forbidden in working process.

7.5.5 Test requirements

No-load test. Under the condition of rated speed, do the positive and reverse rotary for 1 – 2 hours respectively, following are required: stable operation, uniform and small noise, no loose connection, no oil leakage, etc.

Load test. Test should be done under the condition of rated speed and rated power, till the oil temperature is stable. Oil pool temperature rise should not exceed 35 ℃, and bearing temperature rise should not exceed 40 ℃.

7.5.6 Scrub and painting requirements

After passing commissioning test, scrub all components and parts with kerosene, and wash rolling bearing with gasoline. If grease is used in rolling bearing lubrication, grease should be filled into the bearing cavity in right amount (about 1/2 of the cavity volume) before assembly. For the outer surface of housing without cutting, clear the sand away and then coat it with certain color paint.

7.5.7 Lifting requirements

The hook on the base should be used during reducer being moved and lifted. The lifting screw (or lifting eye) on the cover can only be used to lift the cover.

7.5.8 Lubricant requirements

Type, dosage and replacing time of lubricants for transmission components and bearings in the reducer should be determined.

Based on the bearing condition, according to Reference [1], choose oil type and viscosity.

If the same lubricant is used in transmission components and bearings, give priority to the requirements of transmission components.

For multistage transmission, select lubricant according to the average of lubricating oil viscosities for high speed and low speed.

Oil capacity calculation has been described above. Frequency of oil change depends on impurity amount in oil and the degree of its oxidation and contamination, generally half a year is suggested.

7.6 Assembly Drawing Check

After completing the above work, check the assembly drawing according to the following content.

(1) Check the view, if it clearly expresses the working principle of gear reducer and assembly relation, if the projection relation is correct, and if it conforms to the national mechanical drawing standard.

(2) Check the structures of components and parts, if there are any mistakes, if installation, adjustment, maintenance and so on are feasible and convenient.

(3) Check dimensioning, if it is correct, if the locations and dimensions of the important components (such as gear, bearing, shaft, etc.) are consistent with design and calculation, if the dimensions of other related components are coordinated, if fits and precisions are appropriate.

(4) Check if the data in technical characteristic table is correct, if technical conditions are reasonable.

(5) Carefully check component number, check if repetition or omission exists. If the formats and items of title blocks and component list are correct, if the content is correct, etc.

7.7 Error Correction Practice for Assembly Drawing

The purpose of error correction practice is to strengthen the ability of structure analysis, to improve mechanical drawing skills and ability of structure design, and to complete the final design, avoiding major principle or structural errors.

Find following errors in Fig. 7 – 2.

(1) Position of oil outlet is too high, which causes that oil can't be discharged entirely.

(2) The screw head of oil outlet plug should be added with sealing ring.

(3) The locating pin is too short, whose upper and lower ends should be 5 – 10 mm higher than the flange plane.

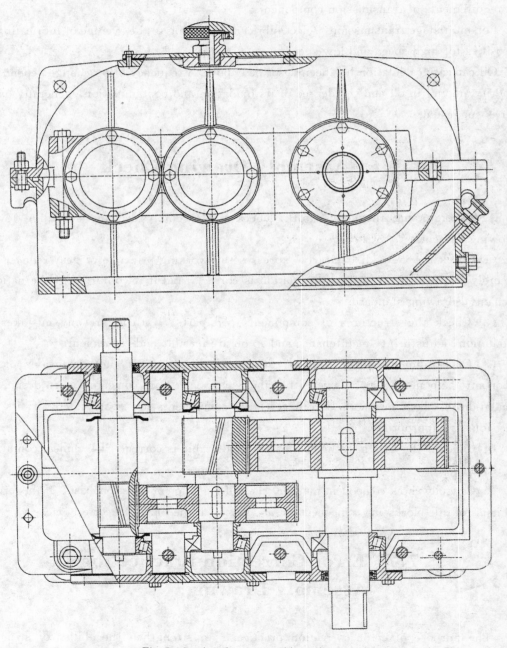

Fig. 7 – 2 A reducer assembly with errors

Chapter 7　Assembly Drawing Design

(4) The spot-facing plane for installation of the bolt connecting upper and lower housings is too small.

(5) The places of connecting bolts near the bearing should have spot-facing planes.

(6) The bore diameter of connecting bolt near the bearing should be greater than the bar diameter.

(7) The section line directions of different partial sections for same component are inconsistent.

(8) There should not be threaded hole on the checking hole lid.

(9) The connecting bolts of bearing covers should not be placed on the split plane.

(10) The bearing cover is solid, should not be split.

(11) If the bearing covers are too close, they should be cut by part, making the distance between them more than 2 mm.

(12) Bearing seat should have draft angle.

(13) Support rib should have draft angle.

(14) The slope of upper housing cover cannot lead the lubricating oil into the oil groove of lower housing.

(15) Gear's addendum is too close to oil bath's bottom.

(16) The direction of oil indicator's scale line is not correct, unable to correctly show the maximum and minimum of oil depth.

(17) The upper surface of observation hole (checking hole) should be 3 – 5 mm higher extending from the external surface of housing, in order to facilitate machining.

(18) The end of connecting bolt on observation hole lid is not expressed completely.

(19) There should be 0.5 mm clearance between oil retainer and bearing seat hole.

(20) Oil retainer is not located, and its position is not correct.

(21) Bearing seat hole should be a through hole.

(22) The end face of bearing seat should extend 5 – 8 mm from the external surface of housing, in order to facilitate machining.

(23) The structure of bearing cover is not correct, not able to generate the circuit of lubricating oil.

(24) The turning of oil groove should not be sharp.

(25) Tooth tip of gear should have chamfer.

(26) Key is too short, and too far away from the assembling end, which is not conducive to assembly.

(27) Keys on the same shaft should be placed on the same generatrix, in order to facilitate machining.

(28) The drawing of key is not correct.

(29) Gears should have foundry fillets.

(30) The segment of shaft fitting with gear is too long, making sleeve's axial locating to gear vain.

(31) The redundant shoulder makes bearing inner ring's locating incorrect.

(32) The structure of shaft is incorrect, which makes it inconvenient for bearing installation.

(33) Shaft end has no chamfer.

(34) Bearing cover should not touch shaft directly.

(35) To facilitate assembly, bearing cover should have back chipping groove.

(36) For the bearing cover's adjusting shim, the part without section is not shown correctly.

(37) The expression of gear meshing zone is not correct.

Chapter 8　Design and Calculation Description

Design and calculation description is the summary of design and calculation process, the theoretical basis of drawing design, and the technical documentation to check if the design satisfies the requirements of production and usage.

8.1　Contents and Requirements

Calculation is the main content of design and calculation description.

Design and calculation description must be written with pen or ball-point pen (pencil is not allowed) on the papers in prescribed format, and bind them together in a volume.

Cover format and writing format are shown in Fig. 8-1.

For binding	Course Design for Machinery Design Major:　　　　 Class:　　　　 Designer:　　　　 Instructor:　　　　 Date:　　　　 (Name of School)

(a)

For binding	Calculation content	Main results

(b)

Fig. 8-1　Description format
(a)Cover format;　(b)Writing format

For the calculation writing, write formulas (including formula sources) first, then put relevant data in, and directly arrive at calculation results without writing calculation process, then indicate the units. Make a brief conclusion to the results, such as "available for the strength requirement" "in the allowable range", etc.

Design and calculation description should include necessary schematic diagram relevant to calculation. For example, in the design and calculation of shaft, structure diagram, force diagram, bending moment diagram, etc. should be included. Other structures, such as gear and sprocket, are not necessary in the description.

In addition to the calculation, technical note should also be included, such as matters needing attention in the process of assembly and disassembly, lubrication methods of transmission parts and rolling bearings, lubricant selection, etc. As the time of course design is limited, the content about the technical note can be simplified.

8.2　Outline

(1) Content (titles and page numbers).
(2) Design task and requirements.
(3) Motor selection.
(4) Kinetic and dynamic parameters of the transmission equipment.
(5) Design calculation for transmission components.
(6) Design calculation for the shafts.
(7) Rolling bearing selection and lifetime checking.
(8) Selection and checking of key joint.
(9) Coupling selection.
(10) Selection of lubricating oil and the form of reducer lubrication and sealing, oil capacity calculation.
(11) References.

PART Ⅱ　　DESIGN INFORMATION AND DATA

Chapter 9　Drawing

9.1　General Regulations

Table 9-1　Dimensions and format of drawing paper (GB/T 14689—93)

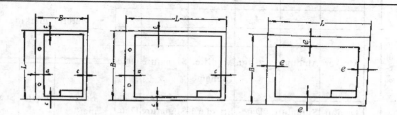

mm

Dimension code	Binding / Basic dimension					No binding / Lengthened dimension					
						Second choice		Third choice			
	A0	A1	A2	A3	A4	Code	$B\times L$	Code	$B\times L$	Code	$B\times L$
Width×Length ($B\times L$)	841×1,189	594×841	420×594	297×420	210×297			A0×2	1,189×1,682	A3×5	420×1,486
						A3×3	420×891	A0×3	1,189×2,523	A3×6	420×1,783
Binding　　a	25					A3×4	420×1,189	A1×3	841×1,783	A3×7	420×2,080
						A4×3	297×630	A1×4	841×2,378	A4×6	297×1,261
c	10			5		A4×4	297×841	A2×3	594×1,261	A4×7	297×1,471
						A4×5	297×1,051	A2×4	594×1,682	A4×8	297×1,682
No binding　　e	20		10					A2×5	594×2,102	A4×9	297×1,892

Note: If there is a binding boundary, a is the binding boundary's width and c is other boundary's width. The dimensions in the table can be increased if necessary. Dimensions of A0, A2 and A4 can be increased by times of 1/8 of A0, dimensions of A1 and A3 can be increased by times of 1/4 of A0's shorter boundary, A0 and A1 are allowed to increase two sides at the same time.

Table 9-2 Scale (GB/T 14690—93)

Same with original structure					1 : 1	
Scale of amplification	2 : 1 $\quad$ 2×10^n : 1	(2.5 : 1) $\quad$ (2.5×10^n : 1)	(4 : 1) $\quad$ (4×10^n : 1)	(5 : 1) $\quad$ (5×10^n : 1)		1×10^n : 1
Scale of reduction	(1 : 1.5) $\quad$ (1 : 1.5×10^n)	1 : 2 $\quad$ 1 : 2×10^n	(1 : 3) $\quad$ (1 : 3×10^n)	(1 : 4) $\quad$ (1 : 4×10^n)	1 : 5 $\quad$ 1 : 5×10^n	(1 : 6) $\quad$ (1 : 6×10^n)	1 : 10 $\quad$ 1 : 1×10^n

Note: n is a positive integer.

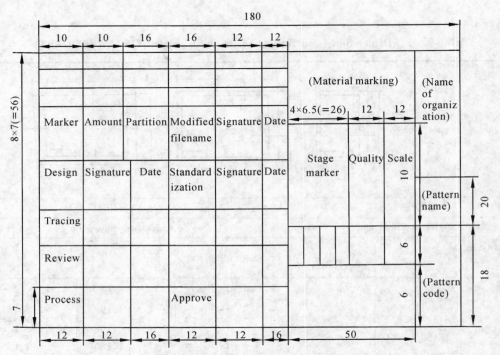

Fig. 9-1 Tile bar format of assemble and component drawing

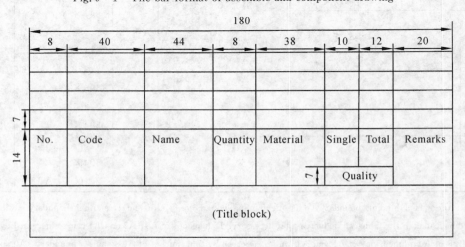

Fig. 9-2 Format of component list

Chapter 9 Drawing

Table 9-3 Name, width, type and general application of drawing line

Name	Width	Type	General application	Name	Width	Type	General application
Solid thick line	b	———	Visible contour line, visible transition line	Double dash and dot line	About $b/3$	—··—··—	Aided parts, contour line, limit position contour line
Fine filling line	About $b/3$	———	Dimension line, dividing line, outlet line, auxiliary line, section line and the line connecting same discontinuous surface	Fine dot and dash line	About $b/3$	—·—·—	Axis line, pitch line, pitch circle and symmetrical line
Dashed line	About $b/3$	- - - - -	Visible contour line, invisible transition line	Wave line	About $b/3$	~~~	Dividing line of visual view and sec-tional view boun-dary line of fracture part

Note: Suggested width of solid thick line is 0.7 mm, suggested width of other line is 0.25 mm.

9.2 Drawing of Commonly Used Components

Table 9-4 Drawing methods of thread and thread fasteners (GB 4459.1—95)

	Explain of drawing methods	Graphic symbol
Drawing methods of external thread and internal thread	The thread's crest is expressed by coarse and bold line, thread root is expressed by fine filling line, both of them should also be drawn on screw's chamfer and fillet. In the projection surface view vertical to thread's axis, 3/4 cycle with fine filling line expresses thread root, and chamfers on shafts and holes shouldn't be drawn (see Fig. (a),(c))	

Continued

	Explain of drawing methods	Graphic symbol
Drawing methods of external thread and internal thread	Effective thread's termination limit should be expressed by coarse and bold line (see Fig. (a)(b)(c)) If the expression of thread end is required, the thread root of screw tail should be drawn by fine filling line which is inclined by 30° with axis (see Fig. (a)) All of the invisible thread's drawing line should be drawn by dashed line (see Fig. (d)) Both of the external thread and the internal thread, their section line should be drawn by coarse and bold line in cutaway view and profile (see Fig. (b)(c)) For the component with conical thread, its thread part's drawing is shown as Fig. (e)(f)	(a) (b) (c) (d) (e) (f)
Drawing methods of the connection of internal and external thread	When the connection of internal and external thread is expressed by cutaway view, the screwing part should be drawn as external thread's drawing method, the others should be drawn by their own drawing method (Fig. (g)(h)) The assembling drawing of wire screw barrel is shown as Fig. (i)(j)	(g) (h) A-A (i) (j)
Drawing methods of threaded fasteners	In assembling drawing, when section plane passes through thread axis, the bolts, nuts and washers are not drawn as section cutting as Fig. (k). Simplified drawing as Fig. (l) can also be used, the head of screw can be drawn as Fig. (m)(n) If threaded hole isn't penetrated in assembling drawing, depth of borehole needn't to be drawn, only the depth of the thread is shown (not including screw tail) (see Fig. (m))	(k) (l) (m) (n)

Chapter 10 Commonly Used Data and General Standard

10.1 Commonly Used Data

Table 10-1 Approximate value and range of transmission ratio of mechanical transmission

Type	Transmission style	Efficiency %	Range of single-stage transmission ratio	
			Maximum	Commonly used
Parallel gear transmission	7 accuracy grade (oil lubrication)	0.98	10	3 – 5
	8 accuracy grade (oil lubrication)	0.97		
	9 accuracy grade (oil lubrication)	0.96		
	exposed drive (grease lubrication)	0.94 – 0.96	15	4 – 6
Bevel gear transmission	7 accuracy grade (oil lubrication)	0.97	6	2 – 3
	8 accuracy grade (oil lubrication)	0.94 – 0.97	6	2 – 3
	Exposed drive (grease lubrication)	0.92 – 0.95	6	4
Belt transmission	V belt transmission	0.95	7	2 – 4
Chain transmission	Roller chain (open type)	0.90 – 0.93	7	2 – 4
	Roller chain (closed form)	0.95 – 0.97		
Worm transmission	Self lock	0.40 – 0.45	Open 100	15 – 16
	Single head	0.70 – 0.75	Closed 80	10 – 40
	Double head	0.75 – 0.82		
	Four head	0.82 – 0.92		
Screw transmission	Lead screw	0.30 – 0.60		
	Rolling screw	0.85 – 0.90		
A pair of rolling bearings	Ball bearing	0.99		
	Roller bearing	0.98		
A pair of sliding bearings	Bad lubrication	0.94		
	Normal lubrication	0.97		
	Liquid friction	0.99		
Coupling	Gear coupling	0.99		
	Elastic coupling	0.99 – 0.995		
Transportation roller		0.96		

10.2 General Standard

Table 10-2 Standard size (diameter, length, height, etc.)

mm

0.1-1.0				10-100						100-1,000					
R		R_a		R			R_a			R			R_a		
R10	R20	R_a 10	R_a 20	R10	R20	R40	R_a 10	R_a 20	R_a 40	R10	R20	R40	R_a 10	R_a 20	R_a 40
0.100	0.100	0.10	0.10								100	100	100	100	100
	0.112		0.11									106			105
0.125	0.125	0.12	0.12	10.0	10.0		10	10			112	112		110	110
	0.140		0.14		11.2			11				118			120
0.160	0.160	0.16	0.16												
	0.180		0.18	12.5	12.5	12.5	12	12	12	125	125	125	125	125	125
0.200	0.200	0.20	0.20		13.2	14.0			13			132			130
	0.224		0.22		14.0	15.0		14	14	140	140			140	140
0.250	0.250	0.25	0.25						15			150			150
	0.280		0.28	16.0	16.0	16.0	16	16	16	160	160	160	160	160	160
0.315	0.315	0.30	0.30			17.0			17			170			170
	0.355					18.0		18	18			180	180		
0.400	0.400	0.40	0.40			19.0			19			190			190
	0.450		0.45	20.0	20.0	20.0	20	20	20	200	200	200	200	200	200
0.500	0.500	0.50	0.50			21.2			21			212			210
	0.560		0.55	22.4	22.4			22	224	224		220	220		
0.630	0.630	0.60	0.60			23.6		24				236			240
	0.710		0.70												
0.800	0.800	0.80	0.80	25.0	25.0	25.0	25	25	25	250	250	250	250	250	250
	0.900		0.90			26.5			26			265			260
1.000	1.000	1.00	1.00		28.0	28.0		28	28			280	280		280	280
1.0-10.0						30.0			30			300			300
R		R_a		31.5	31.5	31.5	32	32	32	315	315	315	320	320	320
R_{10}	R_{20}	R_a	R_a		35.5	33.5			34			335			340
		10	20			35.5		36	36			355	355		360	360

Chapter 10 Commonly Used Data and General Standard

Continued

1.0 – 10.0				10 – 100						100 – 1,000					
R		R_a		R			R_a			R			R_a		
R10	R20	R_a10	R_a20	R10	R20	R40	R_a10	R_a20	R_a40	R10	R20	R40	R_a10	R_a20	R_a40
1.00	1.00	1.0	1.0			37.5		38	38			375			380
	1.12		1.1	40.0	40.0	40.0	40	40	40	400	400	400	400	400	400
1.25	1.25	1.2	1.2			42.5			42			425			420
	1.40		1.4		45.0	45.0		45	45		450	450		450	450
1.60	1.60	1.6	1.6			47.05			48			475			480
	1.80		1.8	50.0	50.0	50.0	50	50	50	500	500	500	500	500	500
2.00	2.00	2.0	2.0			53.0			53			530			530
	2.24		2.2		56.0	56.0		56	56		560	560		560	560
2.50	2.50	2.5	2.5			60.0			60			600			600
	2.80		2.8												
3.15	3.15	3.0	3.0	63.0	63.0	63.0	63	63	63	630	630	630	630	630	630
	3.55		3.5			67.0			67						670
4.00	4.00	4.0	4.0	71.0	71.0		71	71		710	710		710	710	
	4.50		4.5			75.0			75						750
5.00	5.00	5.0	5.0												
	5.60		5.5	80.0	80.0	80.0	80	80	80	800	800	800	800	800	800
6.30	6.30	6.0	6.0			85.0		85			850			850	
	7.10		7.0		90.0	90.0			90		900	900		900	900
8.00	8.00	8.0	8.0			95.0			95			950			950
	9.00		9.0												
10.00	10.0	10.0	10.0	100	100	100	100	100	100	1,000	1,000	1,000	1,000	1,000	1,000

Note: 1. It regulates the common standard size series (diameter, length, height and so on) between 0.01 mm to 2,000 m in mechanical manufacturing industry, suitable for the main dimensions that having interchange ability and series requirements (as installation dimensions, connection dimensions, fit dimensions with tolerance requirement, nominal dimensions that determine series). For other sizes, the series should be used as more as possible. Special standard can be selected for the dimensions with special standard regulations.

2. When select series and single dimension, we should give priority to basic series and single value with bigger tolerance as the priority order of R10, R20 and R40. If value must be rounded, we can select standard dimension in corresponding in R_a series, the priority order is R_a10, R_a20, R_a40.

Table 10-3 Fillet radius and chamfer dimension on fitting surface (GB 6403.4—86)

mm

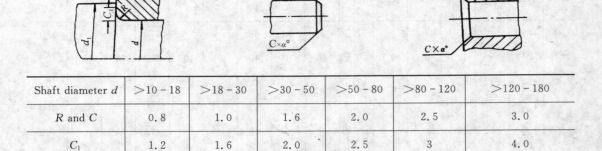

Shaft diameter d	>10-18	>18-30	>30-50	>50-80	>80-120	>120-180
R and C	0.8	1.0	1.6	2.0	2.5	3.0
C_1	1.2	1.6	2.0	2.5	3	4.0

Note: 1. The fillet radius of shaft and bearing seat hole fitting with rolling-contact bearing refers to installation dimensions in Table 13-1 to Table 13-3.
2. a generally is 45°, 30° and 60° can also be used.
3. C_1's value doesn't belong to GB 6403.4—86.
4. $d_1 = d + (3-4)C_1$, and it should be rounded to a standard value.

Table 10-4 Transition fillet radius of free surface of circular component

mm

$D-d$	2	5	8	10	15	20	25	30	35	40	50	55	65	70	90
R	1	2	3	4	5	8	10	12	12	16	16	20	20	25	25

Chapter 11　Thread Fasteners

11.1　Thread Hole

**Table 11-1　Dimensions of through hole and counter bore for fastener
(GB 152.2 - 152.4—88, GB 5277—85)**

mm

Bolt or screw diameter d			4	5	6	8	10	12	14	16	18	20	22	24	27	30
Diameter of through hole d_1		Precision	4.3	5.3	6.4	8.4	10.5	13	15	17	19	21	23	25	28	31
		Medium	4.5	5.5	6.6	9	11	13.5	15.5	17.5	20	22	24	26	30	33
		Crude	4.8	5.8	7	10	12	14.5	16.5	18.5	21	24	26	28	32	35
Counter bore for hexagon head bolt and hexagon head nut	GB 152.4—88	d_2	10	11	13	18	22	26	30	33	36	40	43	48	53	61
		d_3	—	—	—	—	—	16	18	20	22	24	26	28	33	36
		t	Make a plane perpendicular to the shaft axis													
Counter bore for countersunk head	GB 152.2—88	d_2	9.6	10.6	12.8	17.6	20.3	24.4	28.4	32.4	—	40.4	—	—	—	—
		$t\approx$	2.7	2.7	3.3	4.6	5	6	7	8	—	20	—	—	—	—
Counter bore for cylindrical head	GB 152.3—88	d_2	8	10	11	15	18	20	24	26	—	33	—	40	—	48
		d_3	—	—	—	—	—	16	18	20	—	24	—	28	—	36
		t GB 70	4.6	5.7	6.8	9	11	13	15	17.5	—	21.5	—	25.5	—	32
		t GB 65	3.2	4	4.7	6	7	8	9	10.5	—	12.5	—	—	—	—

Note: d_1 in all figures are the same with through hole diameter with medium assembly.

11.2 Bolt

Table 11-2 Hexagon head bolt —level A and level B

—fine pitch level A and level B

mm

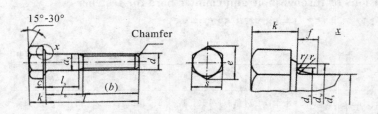

End as GB 2—85 regulations p-pitch

$l_{gmax} = l_{nominal} - b_{reference}$

$l_{smin} = l_{gmax} - 5p$

Marker example

Thread specification d=M12, nominal length l=80 mm, performance grade 8.8, surface oxidation, Level A hexagon head bolt

Bolt GB 5782—2000 M12×80

Thread specification	d GB 5782—86		M6	M8	M10	M12	(M14)	M16	(M18)	M20	(M22)	M24	(M27)	M30
	$d \times p$ GB 5785—86			M8×1	M10×1	M12×1.5	(M14×1.5)	M16×1.5	(M18×1.5)	M20×2	(M20×2)	M24×2	(M27×2)	M30×2
					(M10×1.25)	(M12×1.25)				(M30×1.5)				
b reference	$l<125$		18	22	26	30	34	38	42	46	50	54	60	66
	$125<l<200$			28	32	36	40	44	48	52	56	60	66	72
	$l>200$						53	57	61	65	69	73	79	85
C max			0.5			0.6				0.8				
d_a max			6.8	9.2	11.2	13.7	15.7	17.7	20.2	22.4	24.4	26.4	30.4	33.4
d_s max			6	8	10	12	14	16	18	20	22	24	27	30
d_w min	Product grade	A	8.9	11.6	14.6	16.6	19.6	22.5	25.3	28.2	31.7	33.6	—	—
		B	8.7	11.4	14.4	16.4	19.2	22	24.8	27.7	31.4	33.2	38	42.7
e min	Product grade	A	11.5	14.38	17.77	20.03	23.35	26.75	30.14	33.53	37.72	39.98	—	—
		B	10.89	14.2	17.59	19.85	22.78	26.17	29.56	32.95	37.29	39.55	45.2	50.85
f max			1.4	2			3			4			6	
k nominal			4	5.3	6.4	7.5	8.8	10	11.5	12.5	14	15	17	18.7

Chapter 11 Thread Fasteners

Continued

s max=nominal	10	13	16	18	21	24	27	30	34	36	41	46
r min	0.25	0.4			0.6			0.8	1	0.8	1	
l range	30–60	35–80	40–100	45–120	50–140	55–160	60–180	65–200	70–220	80–240	90–260	90–300
l series nominal	20,25,30,35,40,45,50,55,60,65,70,80,90,100,110,120,130,140,150,160,180, 200,220,240,260,280,300,360,380,400											

Note: 1. Specification in brackets is not suggested.
 2. Product grade: A for $d \leqslant 24$ or $l \leqslant 10d$ (or $l \leqslant 150$ mm); B for $d > 24$ or $l > 10d$ (or $l \leqslant 150$ mm).
 3. Mechanical property level is 8.8.

Table 11–3 Hexagon head bolt – full thread – level A and level B

mm

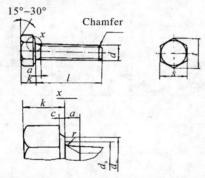

Example: Thread specification $d =$ M12, nominal length $l = 80$ mm, performance level is 8.8, surface oxidation, full thread, Level A Hexagon head bolt
Bolt GB 5783—2000 M12×80

Thread specification d		M5	M6	M8	M10	M12	(M14)	M16	(M18)	M20	(M22)	M24	(M27)	M30
a max		2.4	3	3.75	4.5	5.25	6	6	7.5	7.5	7.5	9	9	10.5
c max		0.5			0.6					0.8				
d_a max		5.7	6.8	9.2	11.2	13.7	15.7	17.7	20.2	22.4	24.4	26.4	30.4	33.4
d_w min	Product grade A	6.9	8.9	11.6	14.6	16.6	19.6	22.5	25.3	28.2	31.7	33.6	—	
	Product grade B	6.7	8.7	11.4	14.4	16.4	19.2	22	24.8	27.7	31.4	33.2	38	
e min	Product grade A	8.79	11.05	14.38	17.77	20.03	23.35	26.75	30.14	33.53	37.72	39.98	—	50.85
	Product grade B	8.63	10.89	14.20	17.59	19.85	22.78	26.17	29.56	32.95	37.29	39.55	45.2	50.85
k nominal		3.5	4	5.3	6.4	7.5	8.8	10	11.5	12.5	14	15	17	18.7
s max		8	10	13	16	18	21	24	27	30	34	36	41	46
r min		0.2	0.25	0.4		0.8				0.8			1	
l range		10–50	12–60	16–80	20–100	25–100	30–140	35–100	35–180	40–100	45–200	40–100	55–200	40–100
l series nominal		6,8,10,12,16,20,25,30,35,40,45,50,(55),60,(65),70–160 (decimal),180,200												

Note: 1. Specification in brackets is not suggested, the thread specification without bracket is product specification.
 2. Note 2 and note 3 are the same with those of Table 11–2.

Table 11-4 Hexagon head bolt for articulation hole—level A and level B(GB 27—88)

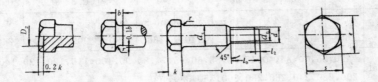

Marker example

Thread specification $d=$ M12, size as the table regulates, nominal length$=$80 mm. Performance level is 8.8, surface is oxidation treated, for level A Hexagon head articulation hole

Bolt GB 27—88 M12×88

When d_s is manufactured as m6, it should be marked with m6

bolt GB 27—88 M12×m6×80

mm

D		6	8	10	12	(14)	16	(18)	20	(22)	24	(27)	30
d_s(h9) max		7	9	11	13	15	17	19	21	23	25	28	32
s max		10	13	16	18	21	24	27	30	34	36	41	46
k nominal		4	5	6	8	8	9	10	11	12	13	15	17
r min		0.25	0.4	0.4	0.6	0.6	0.6	0.6	0.8	0.8	0.8	1.0	1.0
d_p		4	5.5	7	8.5	10	12	13	15	17	18	21	23
l_2		1.5	1.5	2.2	2.2	3	3	3	4	4	4	5	5
e min	A	11.05	14.38	17.77	20.03	23.35	26.75	30.14	33.53	37.72	39.98	—	—
	B	10.89	14.2	17.59	19.85	22.78	26.17	29.56	32.95	37.29	39.55	45.2	50.85
b		2.5	2.5	2.5	2.5	3.5	3.5	3.5	3.5	5	5	5	5
l range		22-65	22-80	30-120	35-180	40-180	45-200	50-200	55-200	60-200	65-200	75-200	80-230
l series		25, (28), 30, (32), 35, (38), 40, 45, 50, (55), 60, 70, (75), 80, 85, 90, (95), 100-260 (decimal), 280, 300											
l_0		12	15	18	22	25	28	30	32	35	38	42	50

Note: 1. Don't use the sizes in brackets as possible as you can.

2. According to operating requirements, the diameter of part without thread (d_s) can be manufactured as m6, u8. For the shank diameter manufactured as m6, its surface roughness is 1.6 μm.

3. The end chamfer of the part without thread (d_s) is 45°, according to manufacturing process requirement, cervical part can be bigger than 45° and less than 1.5p.

11.3 Screw

Table 11-5 Hexagon socket head cap screws

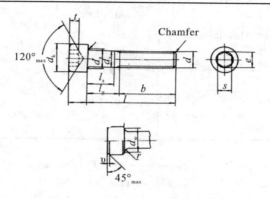

Marker example

Thread specification $d=$ M8, nominal length$=$200 mm. Performance level is 8.8, the surface is oxidation treated.

Screw GB 70—2000 M8×20

mm

Thread specification d		M6	M8	M10	M12	(M14)	M16	M20	M24	M30
p		1	1.25	1.5	1.75	2	2	2.5	3	3.5
b parameter		24	28	32	36	40	44	52	60	72
d_k	Max*	10	13	16	18	21	24	30	36	45
	Max**	10.22	13.27	16.27	18.27	21.33	24.33	30.33	36.39	45.39
d_a max		6.8	9.2	11.2	13.7	15.7	17.7	22.4	26.4	33.4
d_s max		6	8	10	12	14	16	20	24	30
e min		5.72	6.68	9.15	11.43	13.72	16	19.44	21.73	25.15
r min		0.25	0.4	0.4	0.6	0.6	0.6	0.8	0.8	1
k max		6	8	10	12	14	16	20	24	30
s normal		5	6	8	10	12	14	17	19	22
t min		3	4	5	6	7	8	10	12	15.5
v max		0.6	0.8	1	1.2	1.4	1.6	2	2.4	3
d_w min		9.35	12.33	15.33	17.23	20.17	23.17	28.87	34.81	43.61

Continued

l range nominal	10–60	12–80	16–100	20–120	25–140	25–160	30–200	40–200	45–200
When made as full thread $l \leqslant$	30	35	40	45	55	55	65	80	90
l series nominal	10, 12, (14), 16, 20–50(quinary), (55), 60, 65, 70–160(decimal), 180, 200									

Technical condition	Material	Thread tolerance	Grade of the mechanical performance	Tolerance grade of the product	Surface treatment
	Steel	12.9 level is 5g 10g, the others are 6g	8.8, 12.9	A	Oxidation Passivation for zinc plating

Note: 1. M24, M30 is general specification, others are commercial specification.
2. $l_g \min = l_{nominal} - b_{reference}$, $l_s \min = l_g \max - 5p$, p—pitch.
3. Don't use the specification in brackets as possible. * for smooth head ** for knurl head.

Table 11–6 Eye screw (GB 825—88)

mm

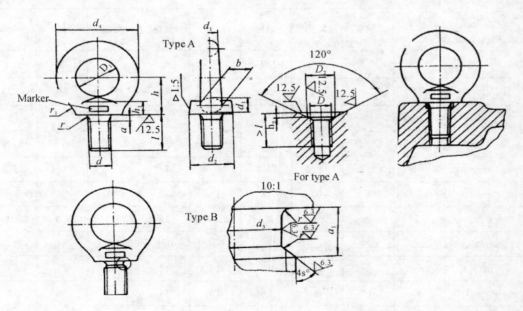

Marker example
Specification is 20 mm, the material is 20 steel, normalized, no surface treated type A eye screw
Screw GB 825—88 M20

Chapter 11　Thread Fasteners

Continued

Thread specification d			M8	M10	M12	M16	M20	M24	M30
d_1	max		9.1	11.1	13.1	15.2	17.4	21.4	25.7
D_1	normal		20	24	28	34	40	48	56
d_2	max		21.1	25.1	29.1	35.2	41.4	49.4	57.7
h_1	max		7	9	11	13	15.1	19.1	23.2
h			18	22	26	31	36	44	53
d_4	reference		36	44	52	62	72	88	104
r_1			4	4	6	6	8	12	15
r	min					1			2
l	normal		16	20	22	28	35	40	45
a_1	max		3.75	4.5	5.25	6	7.5	9	10.5
a	max		2.5	3	3.5	4	5	6	7
b	max		10	12	14	16	19	24	28
d_3	normal(max)		6	7.7	9.4	13	16.4	19.6	25
D_2	normal(min)		13	15	17	22	28	32	38
h_2	normal(min)		2.5	3	3.5	4.5	5	7	8
Maximum lifting load W/kN	Single screw lifting		1.6	2.5	4	6.3	10	16	25
	Double screw lifting		0.8	1.25	2	3.2	5	8	12.5

One-level cylindrical gear reducer					Two-level cylindrical gear reducer					
a	100	160	200	250	315	a	100×140	140×200	180×280	200×280	250×355
W/t	0.026	0.105	0.21	0.40	0.80	W/t	0.10	0.26	0.48	0.68	1.25

Note：1. The relationship between the mass of reducer W and the center distance is for soft gear reducer.

　　　2. The screw is manufactured by 20 or 25 steel, thread tolerance is 8 g.

　　　3. The thread specification d in the table is commercial specification.

Table 11-7 Separating screw

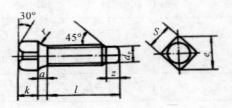

Marker example
Specification d=M10
Nominal length=30 mm
Performance level is 33H
Surface oxidation treated long cylindrical-end set screw
Screw GB 85—88 M10×30

mm

Thread specification d	M5	M6	M8	M10	M12	M16	M20
d_p max	3.5	4	5.5	7	8.5	12	15
l min	6	7.3	9.7	12.2	14.7	20.9	27.1
k normal	5	6	7	8	10	14	18
r min	0.2	0.25	0.4	0.4	0.6	0.6	0.8
S normal	5	6	8	10	12	17	22
z min	2.5	3	4	5	6	8	10
l range	12-13	12-30	14-40	20-50	25-60	25-80	40-100
l series(normal)	8,10,12,(14),16,20,25-55(quinary), 60-100(decimal)						

Technical condition	Material	Grade of the mechanical performance	Thread tolerance	Product grade
	Steel	33H	6g	A
		45H	5g,6g	

Note: The standard is for square-head long cylindrical-end set screw and it can be used as separating screw.

Chapter 11 Thread Fasteners

11.4 Nut

Table 11-8 Type I hex nut- level A and level B
Type I hex nut-fine pitch- level A and level B

mm

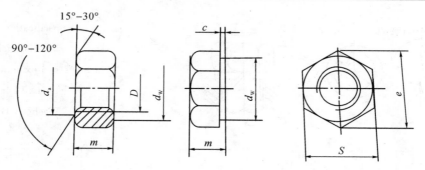

Marker example

Specification $d=$M12, performance grade is 10, no surface treated, level A type I hex nut

Nut GB 6170—2000 M12

Thread specification $D=$M12, performance grade 8, no surface treated, level A type I hex nut

Nut GB 6171—2000 M12×1.5

Thread specification	D	M6	M8	M10	M12	(M14)	M16	(M18)	M20	(M22)	M24	(M27)	M30
	$D×p$		M8×1	M10×1	M12×1.5	M14×1.5	M16×1.5	M18×1.5	M20×2	M22×1.5	M24×2	M27×2	M30×2
c	max	0.5	0.6	0.6	0.6	0.8	0.8	0.8	0.8	0.8	0.8	0.8	0.8
d_a	min	6.75	8	10	12	14	16	18	20	22	24	27	30
d_w	min	8.9	11.6	14.6	16.6	19.6	22.5	24.8	27.7	31.4	33.2	38	42.7
e	min	11.05	14.38	17.77	20.03	23.35	26.75	29.56	32.95	37.29	39.55	45.2	50.85
m	max	5.2	6.8	8.4	10.8	12.8	14.8	15.8	18	19.4	21.5	23.8	25.6
s	max	10	13	16	18	21	24	27	30	34	36	41	46
Technical condition	Material			Grade of the mechanical performance			Thread tolerance			Product grade			
	Steel			6,8,10			6H			A for $D\leqslant16$ B for $D>16$			

Note: Don't use the specification in brackets as possible as you can.

11.5 Washer

Table 11-9 Standard spring washer

mm

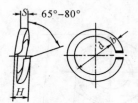 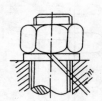

Specification is 16 mm,
Material is 65Mn,
The surface is oxidation treated,
standard spring washer
Washer GB 93—87 16

Specification (major-diameter of the thread)		5	6	8	10	12	(14)	16	(18)	20	(22)	24	(27)	30
d	min	5.1	6.1	8.1	10.2	12.2	14.2	16.2	18.2	20.2	22.5	24.5	27.5	30.5
	max	5.4	6.68	8.68	10.9	12.9	14.9	16.9	19.04	21.04	23.34	25.5	28.5	31.5
$s(b)$	normal	1.3	1.6	2.1	2.6	3.1	3.6	4.1	4.5	5	5.5	6	6.8	7.5
	min	1.2	1.5	2	2.45	2.95	3.4	3.9	4.3	4.8	5.3	5.8	6.5	7.2
	max	1.4	1.7	2.2	2.75	3.25	3.8	4.3	4.7	5.2	5.7	6.2	7.1	7.8
H	min	2.6	3.2	4.2	5.2	6.2	7.2	8.2	9	10	11	12	13.6	15
	max	3.25	4	5.25	6.5	7.75	9	10.25	11.25	12.5	13.75	15	17	18.75
m	$\leqslant$	0.65	0.8	1.05	1.3	1.55	1.8	2.05	2.25	2.5	2.75	3	3.4	3.75

Note: 1. Don't use the specification in brackets as possible as you can.
2. The material is 65Mn, 60Si2Mn, quenched and tempered, hardness is 40-50 HRC.

11.6 Fastening End Cap

Table 11 – 10 Screw fastening end cap and bolt fastening end cap

mm

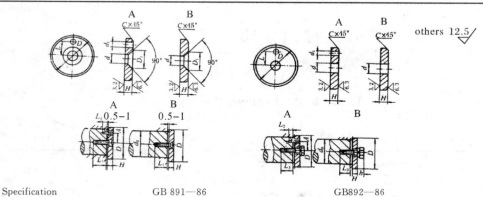

Specification　　　　　　GB 891—86　　　　　　　　GB892—86

$D=$M12, the material is Q235 - A, no surface treated, level A screw fastening end cap
Shield cap GB 892—86 B45
When manufactured as type B, should be marked by B.
Shield cap GB 892—86 B45

Shaft diameter d_0 ≤	Nominal diameter D	H		L		d	d_1	D_1	C	Bolt GB 5783—86	Screw GB 819—87	Cylindrical pin GB 119—86	Washer GB 93—87	Installation size			
		basic size	Limit deviation	basic size	Limit deviation									L_1	L_2	L_3	h
20	28	4		7.5		5.5	2.1	11	0.5	M5×16	M5×12	A2×10	5	14	6	16	5.1
22	30	4		7.5													
25	32	5		10	±0.11	6.6	3.2	13	1	M6×20	M6×16	A3×12	6	18	7	20	6
28	35	5		10													
30	38	5		10													
32	40	5		12													
35	45	5	0 −0.30	12													
40	50	5		12	±0.135												
45	55	6		16		9	4.2	17	1.5	M8×25	M8×20	A4×14	8	22	8	24	8
50	60	6		16													
55	65	6		16													
60	70	6		20													
65	75	6		20													
70	80	6		20	±0.165												
75	90	8	0 −0.36	25		13	5.2	25	2	M12×30	M12×25	A5×16	12	26	10	28	11.5
85	100	8		25													

Note: 1. When shield cap is installed to shaft end with thread hole, the used fastening bolt is allowed to be lengthened.
    2. GB 891—86's mark is same as GB 892—86.

Chapter 12 Key and Pin

12.1 Key

Table 12-1 Common flat key

mm

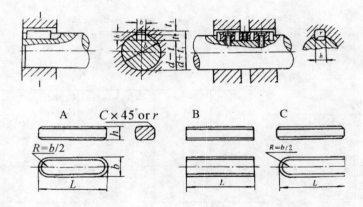

Marker example:

Round-end common flat key (type A) $b=16$ mm, $h=10$ mm, $L=100$ mm; key 16×100 GB 1096—2003

Square-end common flat key (type B) $b=16$ mm, $h=10$ mm, $L=100$ mm; key B16×100 GB 1096—2003

Single round-end common flat key (type C) $b=16$ mm, $h=10$ mm, $L=100$ mm; key C16×100 GB 1096—2003

Chapter 12　Key and Pin

Continued

Shaft	Key	Keyway										
Nominal diameter d	Nominal dimension $b \times h$	Limit deviation of the width b					Depth				Radius r	
		Loose key connection		Normal key connection		Tight key connection	Shaft t		Hub t_1			
		Shaft H9	Hub D10	Shaft N9	Hub JS9	Shaft and hub P9	Nominal Size	Limit deviation	Nominal Size	Limit deviation	Min	Max
>12-17	5×5	+0.030	+0.078	0	±0.015	-0.012	3.0	+0.1	2.3	+0.1	0.16	0.25
>17-22	6×6	0	0.030	+0.30		0.042	3.5	1	2.8	1		
>22-30	8×7	+0.036	+0.098	0	±0.018	-0.015	4.0		3.3			
>30-38	10×8	0	+0.040	-0.036		-0.051	5.0		3.3			
>38-44	12×8						5.0		3.3			
>44-50	14×9	+0.043	+0.120	0	±0.021,5	-0.018	5.5		3.8		0.25	0.40
>50-58	16×10	0	+0.050	-0.043		-0.061	6.0	+0.2	4.3	+0.2		
>58-65	18×11						7.0	0	4.4	0		
>65-75	20×12						7.5		4.9			
>75-85	22×14	+0.052	+0.149	0	±0.026	-0.022	9.0		5.4		0.40	0.60
>85-95	25×14	0	+0.065	-0.052		-0.074	9.0		5.4			
>95-110	28×16						10.0		6.4			
Series of length	14, 16, 18, 20, 22, 25, 28, 32, 36, 4045, 50, 56, 63, 70, 80, 90, 100, 110, 125, 140, 160, 180, 200, 250, 280, 320, 360											

Note:1. On the drawing, keyway depth of shaft is marked as t or $(d-t)$, keyway depth of hub is marked as $(d+t_1)$.

2. The limit deviation of combination size $(d-t)$ and $(d+t_1)$ should be chosen according to relevant limit deviation of t and t_1, but $(d-t)$ should be minus $(-)$.

3. The tolerance of key length L is h14, the tolerance of width b is h9, the tolerance of height h is h11.

4. The surface roughness parameter R of sides of keyway of shaft and hub width b is recommended to be 1.6~3. 2 μm the surface roughness parameter of the bottom surface of keyway of shaft and hub R_a is 6.3 μm.

12.2 Pin

Table 12-2 Cylindrical pin (GB 119—86), taper pin (GB 117—86)

mm

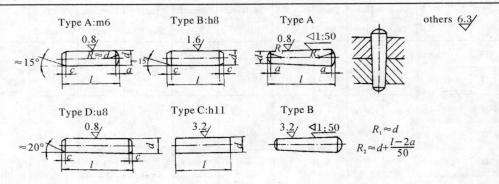

Marker example

Nominal diameter $d=8$ mm, length $l=30$ mm, material is 35 steel, heat treatment hardness is 28-38 HRC, surface oxidation treated type A cylindrical pin   GB 119—86 A8×30

Nominal diameter $d=10$ mm, length $l=60$ mm, material is 35 steel, heat treatment hardness is 28-38HRC, surface oxidation treated type A taper pin GB 117—2000 A10×60

		Nominal	2	3	4	5	6	8	10	12	16	20	25	30
d	Cylindrical pin	Type A Min	2.002	3.002	4.004	5.004	6.004	8.006	10.006	12.007	16.007	20.008	25.008	30.008
		Type A Max	2.008	3.008	4.012	5.012	6.012	7.015	10.015	12.018	16.018	20.021	25.012	30.021
		Type B Min	1.986	2.986	3.982	4.982	5.982	8.978	9.978	11.978	15.973	19.967	24.967	29.967
		Type B Max	2	3	4	5	6	8	10	12	16	20	25	30
		Type C Min	1.94	2.94	3.925	4.925	5.925	7.91	9.91	11.89	15.89	19.87	24.87	29.87
		Type C Max	2	3	4	5	6	8	10	12	16	20	25	30
		Type D Min	2.018	3.018	4.023	5.023	6.023	8.028	10.028	12.033	16.033	20.041	25.048	30.048
		Type D Max	2.032	3.032	4.041	5.041	6.041	8.050	10.050	12.06	16.06	20.074	25.081	30.081
	Taper pin	Min	1.96	2.96	3.95	4.95	5.95	7.94	9.94	11.93	15.93	19.92	24.92	29.92
		max	2	3	4	5	6	8	10	12	16	20	25	30
$a\approx$			0.25	0.40	0.5	0.63	0.80	1.0	1.2	1.6	2.0	2.5	3.0	4.0
$c\approx$			0.35	0.50	0.63	0.80	1.2	1.6	2.0	2.5	3.0	3.5	4.0	5.0
Specification limit of l	Cylindrical pin		6-20	8-28	8-35	10-50	12-60	14-80	16-95	22-140	26-180	35-200	50-200	60-200
	Taper pin		10-35	12-45	14-55	18-60	22-90	22-120	26-160	32-180	40-200	45-200	50-200	55-200
Series nominal			6,8,12,14,16,18,20,22,24,26,28,30,32,35-100 (decimal),120,140,160,180,200											

Note: Material is 34 or 35, heat treatment hardness is 28-38HRC or 38-48HRC.

Chapter 13 Rolling-contact Bearing

Table 13-1 Deep groove ball bearing (GB/T 276—94)

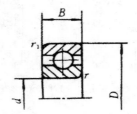 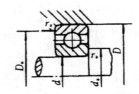 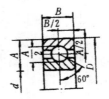

Example: Rolling-contact bearing 6120 GB/T 276—94

A/C_{0r}	e	Y	(radial direction) Equivalent dynamic load	(radial direction) Equivalent static load
0.014	0.19	2.30	When $A/R \leqslant e$,	When $A/R \leqslant 0.8$,
0.028	0.22	1.99	$P = R$	$P_0 = R$
0.056	0.26	1.71	When $A/R > e$,	When $A/R > 0.8$,
0.084	0.28	1.55	$P = 0.56R + YA$	$P_0 = 0.6R + 0.5A$
0.11	0.30	1.45	R—radial load	R—radial load
0.17	0.34	1.31	A—axial load	A—axial load
0.28	0.38	1.15		
0.42	0.42	1.01		
0.56	0.44	1.00		

Type of bearing	Outline dimensions/mm					Installation dimensions/mm			Basic rated load/kN		Speed limit/(r·min^{-1})		Mass ($\approx$)/kg
	d	D	B	r_{min}	r_{1min}	d_{amin}	D_{amax}	r_{amax}	C_r (Dynamic)	C_{0r} (Static)	Grease lubrication	Oil lubrication	
02 series													
6204	20	47	14	1		26	41	1	9.88	6.18	14,000	18,000	0.098
6205	25	52	15	1		31	46	1	10.8	6.95	12,000	16,000	0.121
6206	30	62	16	1		36	56	1	15.0	10.0	9,500	13,000	0.200
6207	35	72	17	1.1		42	65	1	19.8	13.5	8,500	11,000	0.288
6208	40	80	18	1.1		47	73	1	22.8	15.8	8,000	10,000	0.368

Continued

6209	45	85	19	1.1		52	78	1	24.5	17.5	7,000	9,000	0.414
6210	50	90	20	1.1		57	83	1	27.0	19.8	6,700	8,500	0.463
6211	55	100	21	1.5		64	91	1.5	33.5	25.0	6,000	7,500	0.603
6212	60	110	22	1.5		69	101	1.5	36.8	27.8	5,600	7,000	0.780
6213	65	120	23	1.5		74	111	1.5	44.0	34.0	5,000	6,300	0.957
6214	70	125	24	1.5		70	116	1.5	46.8	37.5	4,800	6,000	1.100
6215	75	130	25	1.5	0.5	84	121	1.5	50.8	41.2	4,500	5,600	1.160
6216	80	140	26	2		90	130	2	55.0	44.8	4,300	5,300	1.450
6217	85	150	28	2		95	140	2	64.0	53.2	4,000	5,000	1.780
6218	90	160	30	2		100	150	2	73.8	60.5	3,800	4,800	2.180
6219	95	170	32	2.1		107	158	2.1	84.8	70.5	3,600	4,500	2.620
6220	100	180	34	2.1		112	168	2.1	94.0	79.0	3,400	4,300	3.200
03 series													
6304	20	52	15	1.1		27	45	1	12.2	7.78	13,000	17,000	0.149
6305	25	62	17	1.1		32	55	1	17.2	11.2	10,000	14,000	0.231
6306	30	72	19	1.1		37	65	1	20.8	14.2	9,000	12,000	0.349
6307	35	80	21	1.5		44	71	1.5	25.8	17.8	8,000	10,000	0.455
6308	40	90	23	1.5		48	81	1.5	31.2	22.2	7,000	9,000	0.624
6309	45	100	25	1.5		54	91	1.5	40.8	29.8	6,300	8,000	0.837
6310	50	110	27	2		60	100	2	47.5	35.6	6,000	7,500	1.090
6311	55	120	29	2		65	110	2	55.2	41.8	5,800	6,700	1.355
6312	60	130	31	2.1		72	118	2.1	62.8	48.5	5,600	6,300	1.170
6313	65	140	33	2.1		77	128	2.1	72.2	56.5	4,500	5,600	2.100
6314	70	150	35	2.1		82	138	2.1	80.2	63.2	4,300	5,300	2.550
6315	75	160	37	2.1		87	148	2.1	87.2	71.5	4,000	5,000	3.050
6316	80	170	39	2.1		92	158	2.1	94.5	80.0	3,800	4,800	3.620
6317	85	180	41	3		99	166	2.5	102	89.2	3,600	4,500	4.270
6318	90	190	43	3		104	176	2.5	112	100	3,400	4,300	4.960
6319	95	200	45	3		109	186	2.5	112	112	3,200	4,000	5.720
6320	100	215	47	3		114	201	2.5	132	132	2,800	3,600	7.070

Table 13-2 Angular contact ball bearing (GB/T 292—94)

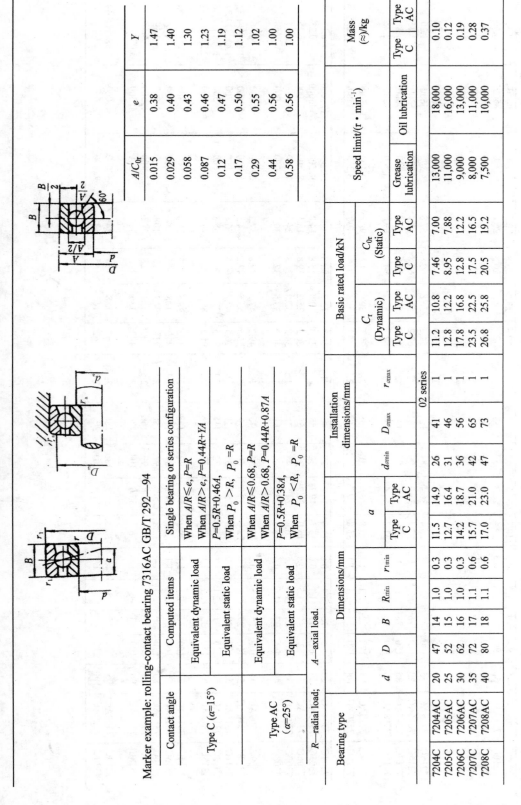

Marker example: rolling-contact bearing 7316AC GB/T 292—94

Contact angle	Computed items	Single bearing or series configuration
Type C ($\alpha=15°$)	Equivalent dynamic load	When $A/R \leq e$, $P=R$ When $A/R > e$, $P=0.44R+YA$
	Equivalent static load	$P_0=0.5R+0.46A$ When $P_0 > R$, $P_0 = R$
Type AC ($\alpha=25°$)	Equivalent dynamic load	When $A/R \leq 0.68$, $P=R$ When $A/R > 0.68$, $P=0.44R+0.87A$
	Equivalent static load	$P_0=0.5R+0.38A$ When $P_0 < R$, $P_0 = R$

R—radial load; A—axial load.

Bearing type		Dimensions/mm				a		Installation dimensions/mm			Basic rated load/kN				Speed limit/(r·min^{-1})		Mass (≈)/kg	
											C_r (Dynamic)		C_{0r} (Static)					
	d	D	B	R_{min}	r_{1min}	Type C	Type AC	d_{amin}	D_{amax}	r_{amax}	Type C	Type AC	Type C	Type AC	Grease lubrication	Oil lubrication	Type C	Type AC
										02 series								
7204C 7204AC	20	47	14	1.0	0.3	11.5	14.9	26	41	1	11.2	10.8	7.46	7.00	13,000	18,000	0.10	0.10
7205C 7205AC	25	52	15	1.0	0.3	12.7	16.4	31	46	1	12.8	12.2	8.95	7.88	11,000	16,000	0.12	0.12
7206C 7206AC	30	62	16	1.0	0.3	14.2	18.7	36	56	1	17.8	16.8	12.8	12.2	9,000	13,000	0.19	0.19
7207C 7207AC	35	72	17	1.1	0.6	15.7	21.0	42	65	1	23.5	22.5	17.5	16.5	8,000	11,000	0.28	0.28
7208C 7208AC	40	80	18	1.1	0.6	17.0	23.0	47	73	1	26.8	25.8	20.5	19.2	7,500	10,000	0.37	0.37

A/C_{0r}	e	Y
0.015	0.38	1.47
0.029	0.40	1.40
0.058	0.43	1.30
0.087	0.46	1.23
0.12	0.47	1.19
0.17	0.50	1.12
0.29	0.55	1.02
0.44	0.56	1.00
0.58	0.56	1.00

Continued

7209C	7209AC	45	85	19	1.1	0.6	18.2	24.7	52	78	1	29.8	28.2	23.8	22.5	6,700	9000	0.41
7210C	7210AC	50	90	20	1.1	0.6	19.4	26.3	57	83	1	32.8	31.5	26.8	25.2	6,300	8500	0.46
7211C	7211AC	55	100	21	1.1	0.6	20.9	28.6	64	91	1.5	40.8	38.8	33.8	31.8	5,600	7500	0.61
7212C	7212AC	60	110	22	1.5	0.6	22.4	30.8	69	101	1.5	44.8	42.8	37.8	35.5	5,300	7000	0.80
7213C	7213AC	65	120	23	1.5	0.6	24.2	33.5	74	111	1.5	53.8	51.2	46.0	43.2	4,800	6300	1.00
7214C	7214AC	70	125	24	1.5	0.6	25.3	35.1	79	116	1.5	56.0	53.2	49.2	46.2	4,500	6000	1.10
7215C	7215AC	75	130	25	2	1	26.4	36.4	84	121	1.5	54.2	50.8	60.8	57.8	4,300	5600	1.20
7216C	7216AC	80	140	26	2	1	27.7	38.9	90	131	2	63.2	59.2	68.8	65.5	4,000	5300	1.45
7217C	7217AC	85	150	28	2	1	29.9	41.6	95	140	2	69.8	65.5	76.8	72.8	3,800	5000	1.80
7218C	7218AC	90	160	30	2	1	31.7	44.4	100	150	2	87.8	82.2	94.2	89.8	3,600	4800	2.25
7219C	7219AC	95	170	32	2.1	1.1	33.8	46.9	107	158	2.1	95.5	89.9	102	98.8	3,400	4500	2.70
7220C	7220AC	100	180	34	2.1	1.1	35.8	49.7	112	168	2.1	140	108	115	100	3,200	4300	3.25

03 series

7304C	7304AC	20	52	15	1.1	0.6	11.3	16.3	27	45	1	14.2	13.8	9.68	9.0	12,000	17,000	0.15
7305C	7305AC	25	62	17	1.1	0.6	13.1	19.1	32	55	1	21.4	20.8	15.8	14.8	9,500	14,000	0.23
7306C	7306AC	30	72	19	1.5	0.6	15.0	22.2	37	65	1	26.2	25.2	19.8	18.5	8,500	12,000	0.35
7307C	7307AC	35	80	21	1.5	0.6	16.6	24.5	44	71	1.5	34.2	32.8	26.8	24.8	7,500	10,000	0.47
7308C	7308AC	40	90	23	1.5	0.6	18.5	27.5	49	81	1.5	40.2	38.5	32.8	30.5	6,700	9,000	0.66
7309C	7309AC	45	100	25	1.5	0.6	20.2	30.2	54	91	1.5	49.2	47.5	39.8	37.2	6,000	8,000	0.86
7310C	7310AC	50	110	27	2	1.0	22.0	33.0	60	99	2	55.5	53.5	47.2	44.5	5,600	7,500	1.08	1.32
7311C	7311AC	55	120	29	2	1.0	23.8	35.8	65	110	2	70.5	67.2	60.5	56.8	5,000	6,700	1.42	1.71
7312C	7312AC	60	130	31	2.1	1.1	25.6	38.7	72	118	2.1	80.5	77.8	70.2	65.8	4,800	6,300	1.70	2.06
7313C	7313AC	65	140	33	2.1	1.1	27.4	41.5	77	128	2.1	91.5	89.8	80.5	70.5	4,300	5,600	2.23	2.57
7314C	7314AC	70	150	35	2.1	1.1	29.2	44.3	82	138	2.1	102	98.5	91.5	86.0	4,000	5,300	2.67	3.06
7315C	7315AC	75	160	37	2.1	1.1	31	47.2	87	148	2.1	112	108	105	97	3,800	5,000	3.56
7316C	7316AC	80	170	39	2.1	1.1	32.8	50	92	158	2.1	118	115	118	108	3,600	4,800	3.59
7317C	7317AC	85	180	41	3	1.1	34.6	52.8	99	166	2.5	132	122	128	122	3,400	4,500	4.38
7318C	7318AC	90	190	43	3	1.1	36.4	55.6	104	176	2.5	142	135	142	135	3,200	4,300	5.02	5.17
7319C	7319AC	95	200	45	3	1.1	38.2	58.5	186	186	2.5	152	158	158	148	3,000	4,000	5.98
7320C	7320AC	100	215	47	3	1.1	40.2	61.9	201	201	2.5	162	165	165	178	2,600	3,600	7.20	6.69

Chapter 13 Rolling – contact Bearing

Table 13-3 Tapered roller bearing

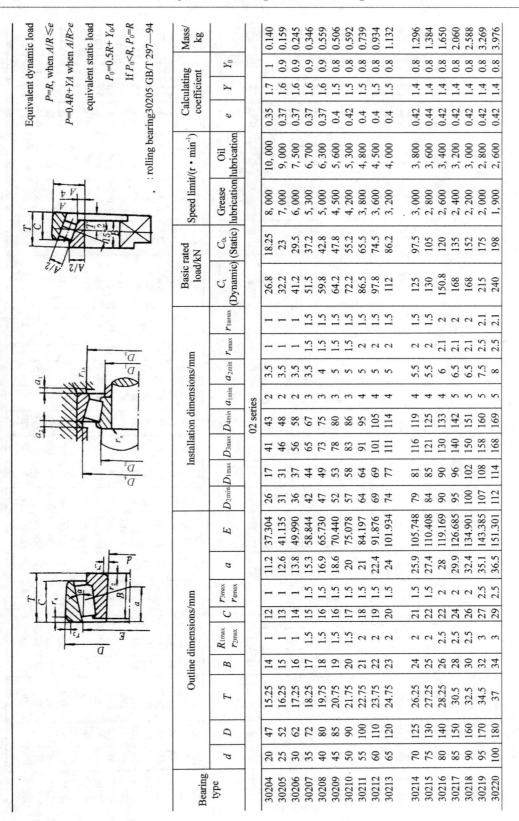

Equivalent dynamic load
$P=R$, when $A/R \leq e$
$P=0.4R+YA$ when $A/R>e$
equivalent static load
$P_0=0.5R+Y_0A$
If $P_0<R$, $P_0=R$

rolling bearing 30205 GB/T 297—94

Bearing type	Outline dimensions/mm									Installation dimensions/mm									Basic rated load/kN		Speed limit/(r·min⁻¹)		Calculating coefficient			Mass/kg
	d	D	T	B	C	R_{1max} r_{2max}	r_{3max} r_{4max}	a	E	D_{2min}	D_{1max}	D_{3max}	D_{4min}	a_{1min}	a_{2min}	r_{amax}	r_{1amax}	C_r (Dynamic)	C_{0r} (Static)	Grease lubrication	Oil lubrication	e	Y	Y_0	
02 series																										
30204	20	47	15.25	14	12	1	1	11.2	37.304	26	17	41	43	2	3.5	1	1	26.8	18.25	8,000	10,000	0.35	1.7	1	0.140
30205	25	52	16.25	15	13	1	1	12.6	41.135	31	31	46	48	2	3.5	1	1	32.2	23	7,000	9,000	0.37	1.6	0.9	0.159
30206	30	62	17.25	16	14	1	1	13.8	49.990	36	37	56	58	2	3.5	1	1	41.2	29.5	6,000	7,500	0.37	1.6	0.9	0.245
30207	35	72	18.25	17	15	1.5	1.5	15.3	58.844	42	44	65	67	3	3.5	1.5	1.5	51.5	37.2	5,300	6,700	0.37	1.6	0.9	0.346
30208	40	80	19.75	18	16	1.5	1.5	16.9	65.730	47	49	73	75	3	4	1.5	1.5	59.8	42.8	5,000	6,300	0.37	1.6	0.9	0.559
30209	45	85	20.75	19	16	1.5	1.5	18.6	70.440	52	53	78	80	3	5	1.5	1.5	64.2	47.8	4,500	5,600	0.4	1.5	0.8	0.506
30210	50	90	21.75	20	17	1.5	1.5	20	75.078	57	58	83	86	3	5	1.5	1.5	72.2	55.2	4,200	5,300	0.42	1.5	0.8	0.592
30211	55	100	22.75	21	18	2	1.5	21	84.197	64	64	91	95	4	5	2	1.5	86.5	65.5	3,800	4,800	0.4	1.5	0.8	0.739
30212	60	110	23.75	22	19	2	1.5	22.4	91.876	69	69	101	105	4	5	2	1.5	97.8	74.5	3,600	4,500	0.4	1.5	0.8	0.934
30213	65	120	24.75	23	20	2	1.5	24	101.934	74	77	111	114	4	5	2	1.5	112	86.2	3,200	4,000	0.4	1.5	0.8	1.132
30214	70	125	26.25	24	21	2	1.5	25.9	105.748	79	81	116	119	4	5.5	2	1.5	125	97.5	3,000	3,800	0.42	1.4	0.8	1.296
30215	75	130	27.25	25	22	2	1.5	27.4	110.408	84	85	121	125	4	5.5	2	1.5	130	105	2,800	3,600	0.44	1.4	0.8	1.384
30216	80	140	28.25	26	22	2.5	2	28	119.169	90	90	130	133	4	6	2.1	2	150.8	120	2,600	3,400	0.42	1.4	0.8	1.650
30217	85	150	30.5	28	24	2.5	2	29.9	126.685	95	96	140	142	5	6.5	2.1	2	168	135	2,400	3,200	0.42	1.4	0.8	2.060
30218	90	160	32.5	30	26	2.5	2	32.4	134.901	100	102	150	151	5	6.5	2.1	2	168	152	2,200	3,000	0.42	1.4	0.8	2.588
30219	95	170	34.5	32	27	3	2.5	35.1	143.385	107	108	158	160	5	7.5	2.5	2.1	215	175	2,000	2,800	0.42	1.4	0.8	3.269
30220	100	180	37	34	29	3	2.5	36.5	151.301	112	114	168	169	5	8	2.5	2.1	240	198	1,900	2,600	0.42	1.4	0.8	3.976

· 85 ·

Continued

03 series

型号	d	D	T	B	r	C	r₁	a	Cr	C0r											ng(grease)	n0(oil)	e	Y	Y0	factor
30304	20	52	16.25	15	1.5	13	1.5	11	41.318	27	28	45	48	3	3.5	1.5	1.5	31.5	20.8	7,500	9,500	0.3	2	1.1	0.168
30305	25	62	18.25	17	1.5	15	1.5	13	50.637	32	34	55	58	3	3.5	1.5	1.5	44.8	30	6,300	8,000	0.3	2	1.1	0.250
30306	30	72	20.75	19	1.5	16	1.5	15	58.278	37	40	65	66	3	5	1.5	1.5	55.8	38.5	5,600	7,000	0.31	1.9	1	0.408
30307	35	80	22.75	21	2	18	1.5	17	65.769	44	45	71	74	3	5	2	1.5	62.2	44.5	5,000	6,300	0.31	1.9	1	0.530
30308	40	90	25.25	23	2	20	1.5	19.5	72.703	49	52	81	84	3	5.5	2	1.5	86.2	63.8	4,500	5,600	0.35	1.7	1	0.761
30309	45	100	27.25	25	2	22	1.5	21.5	81.780	54	59	91	94	3	5.5	2	1.5	104	78.1	4,000	4,800	0.35	1.7	1	1.066
30310	50	110	29.25	27	2.5	23	2	23	90.633	60	65	100	103	4	6.5	2.1	2	122	92.5	3,800	5,000	0.35	1.7	1	1.250
30311	55	120	31.5	29	2.5	25	2	25	99.146	65	70	110	112	4	6.5	2.1	2	145	112	3,400	4,300	0.35	1.7	1	1.528
30312	60	130	33.5	31	3	26	2.5	26.5	107.769	72	76	118	121	5	7.5	2.5	2.1	185	125	3,200	4,300	0.35	1.7	1	1.940
30313	65	140	36	33	3	28	2.5	29	125.244	87	83	128	131	5	8	2.5	2.1	162	142	2,800	4,000	0.35	1.7	1	2.629
30314	70	150	38	35	3	30	2.8	30.6	116.846	77	89	138	141	5	8	2.5	2.1	208	162	2,600	3,400	0.35	1.7	1	3.170
30315	75	160	40	37	3	31	2.5	32	134.097	82	95	148	160	5	9	2.5	2.1	238	188	2,400	3,200	0.35	1.7	1	3.542
30316	80	170	42.5	39	3	33	2.5	34	143.174	92	102	158	150	5	9.5	2.5	2.1	262	208	2,200	3,000	0.35	1.7	1	4.486
30317	85	180	44.5	41	4	34	3	36	150.433	99	113	166	168	6	10.5	3	2.5	288	228	2,000	2,800	0.35	1.7	1	5.305
30318	90	190	46.5	43	4	36	3	37.5	159.061	104	107	176	178	6	10.5	3	2.5	322	260	1,900	2,600	0.35	1.7	1	6.144
30319	95	200	49.5	45	4	38	3	40	165.861	109	118	186	185	6	11.5	3	2.5	348	282	1,800	2,400	0.35	1.7	0.8	7.130
30320	100	215	51.5	47	4	39	3	42	178.578	114	127	201	199	6	12.5	3	2.5	382	310	1,600	2,000	0.35	1.7	1	8.690

22 series

型号	d	D	T	B	r	C	r₁	a	Cr	C0r											ng	n0	e	Y	Y0	factor
32206	30	62	21.25	20	1	17	1	15.4	48.982	36	36	56	58	3	4.5	1	1	49.2	37.2	6,000	7,500	0.37	1.6	0.9	0.285
32207	35	72	24.25	23	1.5	19	1.5	17.6	57.087	42	42	65	68	3	5.5	1.5	1.5	67.5	52.5	5,300	6,700	0.37	1.6	0.9	0.488
32208	40	80	24.75	23	1.5	19	1.5	19	57.087	47	48	73	75	3	6	1.5	1.5	74.2	56.8	5,000	6,300	0.4	1.6	0.9	0.577
32209	45	85	24.75	23	1.5	19	1.5	20	69.610	52	53	78	86	3	6	1.5	1.5	79.5	62.8	4,500	5,600	0.4	1.5	0.8	0.577
32210	50	90	24.75	23	1.5	19	1.5	21	74.226	57	57	83	96	3	6	1.5	1.5	84.8	68	3,800	5,300	0.42	1.4	0.8	0.618
32211	55	100	26.75	25	2	21	1.5	22.5	82.837	64	62	91	81	4	6	2	1.5	102	81.5	4,300	4,800	0.4	1.5	0.8	0.915
32212	60	110	29.75	28	2	24	1.5	24.9	99.484	69	68	101	105	4	6	2	1.5	125	102	3,600	4,500	0.4	1.5	0.8	1.197
32213	65	120	33.25	31	2	27	1.5	27.2	90.236	74	75	111	115	4	6	2	1.5	152	125	3,000	4,000	0.4	1.5	0.8	1.580
32214	70	125	32.75	31	2	27	1.5	28.6	103.765	79	79	116	120	4	6	2	1.5	158	135	2,800	3,800	0.42	1.4	0.8	1.620
32215	75	130	33.25	31	2	27	1.5	30.2	108.932	84	84	121	126	4	6.5	2	1.5	160	135	2,600	3,600	0.44	1.4	0.8	1.765
32216	80	140	35.25	33	2.5	28	2	1.3	124.970	90	89	130	135	5	7.5	2.1	2	188	158	2,400	3,400	0.42	1.4	0.8	2.670
32217	85	150	38.5	36	2.5	30	2	34	132.615	95	101	140	143	5	8.5	2.1	2	215	185	200	3,200	0.42	1.4	0.8	2.162
32218	90	160	42.5	40	2.5	34	2	36.7	140.259	100	95	150	153	5	8.5	2.5	2.1	258	225	1,900	2,800	0.42	1.4	0.8	3.265
32219	95	170	45.5	43	3	37	2.5	39	148.184	107	106	158	163	5	8.5	2.5	2.5	285	255	2,000	3,000	0.42	1.4	0.8	4.216
32220	100	180	49	46	3	39	2.5	41.8	117.466	112	113	168	172	5	10	2.5	2.1	322	292	3,200	2,600	0.42	1.4	0.8	5.230

Chapter 13 Rolling-contact Bearing

Table 13-4 Fitting between rolling bearing and shaft or bearing seat hole

Bearing type / Load properties	Deep groove ball bearing / Angular contact ball bearing	Tapered roller bearing	Fit code	
P/C	Nominal diameter of the bearing/mm		Shaft	Bearing seat hole
$\leqslant 0.07$	20 - 100 $>$100 - 200	$\leqslant 40$ $>$40 - 140	js6, j6 k6	H7
$>0.07 - 0.15$	20 - 100 $>$100 - 140 $>$140 - 200	$\leqslant 40$ $>$40 - 100 $>$100 - 140	k5, k6 m5, m6 m6	H7
>0.15		$>$50~140	n6	H7

Note: P is equivalent dynamic load, C is basic rated dynamic load.

Chapter 14 Coupling

Table 14-1 HL type flexible pin coupling

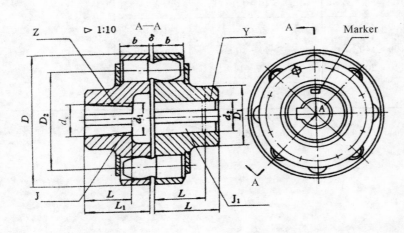

Marker example 1
HL6 coupling 65×142
GB/T 5014—2003
Driving shaft $d_1 = 65$ mm,
Y- type shaft hole $L = 142$ mm
A- type keyway
Driven shaft $d_2 = 65$ mm,
Y- type shaft hole $L = 142$ mm
A- type keyway
Marker example 2
HL7 coupling $\dfrac{ZC\ 75 \times 107}{JB\ 70 \times 107}$
GB/T 5014—2003
Driving shaft $d_z = 75$ mm,
Z-type shaft hole $L_1 = 107$ mm
C-type keyway
Driven shaft $d_2 = 70$ mm,
J-type shaft hole $L_1 = 107$ mm
B-type keyway

Type	Nominal torque T_n/(N·m)	Allowable rotating speed $[n]$/(r·min^{-1})		Hole diameter d_1, d_2, d_z		Hole length			D	D_1	D_2	b	s	Moment of inertia/ (kg·m^2)	Mass/ kg
						Y	J, J$_1$	Z							
				Steel	Iron	L_1	L	L_1							
		Steel	Iron				mm								
HL1	160	7,100		12	12	32	37	32	90	40	65	20	2.5	0.006,4	2
				14	14										
				16	16	42	30	42							
				18	18										
				19	19										
				20	20	52	38	52							
				22	22										
				24											
HL2	315	5,600		20	20	52	38	52	120	55	90	28	2.5	0.253	5
				22	22										
				24	24										
				25	25	62	44	62							
				28	28										
				30	30										
				32	32	82	60	82							
				35											

Chapter 14 Coupling

Continued

HL3	630	5,000		30 32 35 38	30 32 35 38	82	60	82	160	75	125	36	2.5	0.6	8
				40 42 45 48	40 42 —	—	112	84	112						
HL4	1,250	4,000	2,800	40 42 45 48 50 55 56	40 42 45 48 50 55 56	112	84	112	195	100	150	45	3	3.4	22
				60 63	— —	142	107	142							
HL5	2,000	3,500	2,500	50 55 56	50 55 56	112	84	112	220	120	170	45	3	5.4	30
				60 63 65	60 63 65	142	107	142							
				70 71 75	70 —										
HL6	3,150	2,800	2,100	60 63 65 70 71 75	60 63 65 70 71 75	142	107	142	280	140	220	56	4	15.6	53
				80 85	80 —	172	132	172							
HL7	6,300	2,240	1,700	70 71 75	70 71 75	142	107	142	320	200	250	56	5	41.1	58
				80 85 90 95	80 85 90 95	172	132	172							
				100 110	100 —	212	167	212							

Note: There is no D_2 in original standard, the value in the table is reference dimension.

Table 14-2 HL type flexible pin coupling

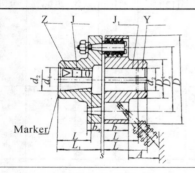

Marker example 1 TL6 coupling 40×112 GB 4323—2002

Driving shaft $d_1=40$ mm, Y-type hole $L=112$ mm, A-type keyway

Driven shaft $d_2=40$ mm, Y-type hole $L=112$ mm, A-type keyway

Marker example 2 TL7 coupling $\dfrac{ZC\ 16\times 30}{JB\ 18\times 30}$ GB 4323—2002

Driving shaft $d_z=16$ mm, Z-type hole $L_1=30$ mm, C-type keyway

Driven shaft $d_2=18$ mm, J-type hole $L_1=30$ mm, B-type keyway

Continued

Type	Nominal torque T_n /(N·m)	Allowable rotating speed $[n]$/(r·min^{-1})		Hole diameter d_1, d_2, d_z		Hole length			D	D_1	D_2	b	b_1	s	A	Moment of inertia/ (kg·m^2)	Mass/ kg
						Y	J, J$_1$, Z										
		Steel	Iron	Steel	Iron	L_1	L	L_1	mm								
TL1	6.3	6,600	8,800	9	9	20	—	—	71	22	45	16	10	3	18	0.000,4	0.7
				10	10	25	22	25									
				11	11												
				12	12	32	27	32									
					14												
TL2	16	5,500	7,600	12	12				80	30	53					0.001	1.0
				14	14												
				16	16	42	30	42									
				—	18												
				—	19												
				—	20	52	38	52									
TL3	31.5	4,700	6,300	16	16	42	30	42	95	35	63	23	15	4	35	0.04	2.2
				18	18												
				19	19												
				20	20	52	38	52									
					22												
TL4	63	4,200	5,700	20	20				106	42	76					0.011	3.2
				22	22												
				24	24												
				—	25	62	44	62									
				—	28												
TL5	125	3,600	4,600	25	25				130	56	90	38	17	5	45	0.026	5.5
				28	28												
				30	30	82	60	82									
				32	32												
				—	35												
TL6	250	3,300	3,800	32	32				160	71	112					0.06	9.6
				35	35												
				38	38												
				40	40	112	84	112									
					42												
TL7	500	2,800	3,600	40	40	112	84	112	190	80						0.06	15.7
				42	42												
				45	45												
					48												

Chapter 14 Coupling

Continued

TL8	710	2,400	3,000	45	45	112	84	112	224	95	48	19	6	65	0.13	24
				48	48											
				50	50											
				55	55											
				—	56											
				—												
				—	60	142	107	142								
				—	63											
TL9	1,000	2,100	2,850	50	50	112	84	112	250	110					0.20	31
				55	55											
				56	56											
				60	60	142	107	142								
				63	63											
				—	65											
				—	70											
				—	71											
				—												
TL10	2,000	1,700	2,300	63	60				315	150	58	22	8	80	0.64	60.2
				65	63											
				70	70											
				71	71											
				75	75											
				80	80	172	132	172								
				85	85											
				—	90											
				—	95											

Note: There is no D_2 in original standard, the value in the table is reference dimension.

Table 14-3 ML plum-blossom type elastic coupling

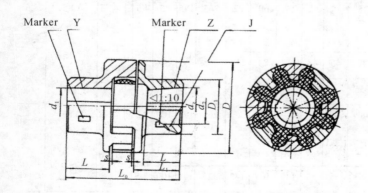

Marker example
ML3 plum-blossom elastic coupling
Hardness of elastic components a
Driving shaft, Z-type shaft hole,
C-type keyway $d_1 = 30$ mm, $L = 60$ mm
Driven shaft Y-type shaft hole,
B-type keyway $d_2 = 25$ mm, $L = 62$ mm
ML3 coupling $\frac{ZC\ 30 \times 60}{YB\ 25 \times 62}$ MT3a
GB 5272—2002

Continued

Type	Nominal torque T_n/(N·m) Elastic component hardness Type A			Allowable rotating speed $[n]$/(r·min⁻¹)		Hole diameter d_1, d_2, d_3			Hole length Y (L)	J (L)	J$_1$ (L_1)	L_{0max}	D	D_1	s	Elastic body type	Moment of inertia (kg·m²)	Mass/kg
	a ≥75	b ≥85	c ≥94	Iron	Steel							mm						
ML1	16	25	45	11,500	15,300	12	14		32	37	32	80	50	40	2	MT1 -a -b -c	0.014	0.66
						16	18	19	42	30	42	100						
						20	22	24	52	38	52	120						
ML2	63	100	200	8,200	10,900	20	22	24				127	70	50	2.5	MT2 -a -b -c	0.075	1.55
						25	28		62	44	62	147						
						30	32		82	60	82	187						
ML3	90	140	280	6,700	9,000	22	24		52	38	52	128	85	60	3	MT3 -a -b -c	0.178	2.5
						25	28		62	44	62	148						
						30	32	35	38	82	60	82	188					
ML4	140	250	400	5,500	7,300	25	28		62	44	62	151	105	65	3.5	MT4 -a -b -c	0.412	4.3
						30	32	35*	38	82	60	82	191					
						40	42		112	84	112	251						
ML5	250	400	710	4,600	6,100	30	32	35	38	82	60	82	197	125	75	4	MT5 -a -b -c	0.73	6.2
						40	42	45	48	112	84	112	257					
ML6	400	630	1,120	4,000	5,300	35*	38*		82	60	82	203	145	85	4.5	MT6 -a -b -c	1.85	8.6
						40*	42*	45	48	112	84	112	263					
						50	55											
ML7	710	1,120	2,240	3,400	4,500	45*	48*					265	175	100	5.5	MT7 -a -b -c	3.86	14
						50	55											
						60	63	65	142	107	142	325	170	100	5.5			

Chapter 14 Coupling

Continued

ML8	1,120	1,800	3,550	2,900	3,800	50*	55*		112	84	112	272	200	120	6.5	-a MT8-b -c	9.22	25.7
						60	63	65	70	142	107	142	232				
						60*	63*	65	70								
						71	75										
ML9	1,800	2,800	5,600	2,500	3,300	60*	63*	65	70					230	150	7.5	-a MT9 -b -c	18.95	41
						71	75										
						80	85	90	95	172	132	172	394				
ML10	2,800	4,500	9,000	2,200	2,900	70*	71*	75*		142	107	142	344	260	180	8.5	-a MT10-b -c	39.68	59
						80*	85*	90	95	172	132	172	404				
						100	110			212	167	212	484				

Note: 1. The diameter of shaft hole with * can be used for Z-type hole.

2. a, b, c is elastic body hardness code.

Chapter 15 Lubrication and Sealing

15.1 Lubricant

Table 15-1 Properties and application of commonly used lubricating oil

Name	Code	Kinematic viscosity			Viscosity index is not less than	Flash point (℃) is not less than	Pour point (℃) is not more than	Main application
		40℃	50℃	100℃				
L-AN oil for total loss system (from GB 443—89)	5	4.14-5.06				80	-5	Main for total loss lubrication system that have no special requirements, not suit for circulating lubrication system
	7	6.12-7.48				110		
	10	9-11				130		
	15	13.5-16.5				150		
	22	19.8-24.2				150		
	32	28.8-35.2				150		
	46	41.4-50.6				160		
	68	61.2-74.8				160		
	100	90-110				180		
	150	130-165				180		
Worm gear oill (from SH 0094—91)	220	198-242	108-129					For the lubrication of worm transmission
	320	288-352	151-182					
	460	414-506	210-252					
	680	612-748	300-360					
	1000	9 00-1,100	425-509					

Chapter 15 Lubrication and Sealing

Continued

Name	Code	Kinematic viscosity			Viscosity index is not less than	Flash point (℃) is not less than	Pour point (℃) is not more than	Main application
		40℃	50℃	100℃				
Antioxygen antirust industry gear oil (from SY 1172—80)	68	61.2 – 74.8	37.1 – 44.4		90	170	−8	For normal gear, gear surface stress should be less than (300 – 500)MPa
	100	90 – 110	52.4 – 63.0					
	150	135 – 165	75.9 – 91.2					
	220	198 – 242	108 – 129			200		
	320	288 – 352	151 – 182					
	460	414 – 506	210 – 252					
	680	612 – 748	300 – 360					
	1000	900 – 1100	466 – 560					
	1500	1350 – 1650	676 – 812					
Middle duty industry gear oil (from GB 5903—86)	68	61.2 – 74.8	31.7 – 44.4		90	180	−8	For middle duty gear or low duty gear, whose gear surface stress is (500 – 1,000) MPa, such as the gear lubrication of chemical, metallurgical, mine and so on
	100	90 – 110	52.4 – 63.0					
	150	135 – 165	75.9 – 91.2			200		
	220	198 – 242	108 – 129					
	320	288 – 352	151 – 182					
	460	414 – 506	210 – 252					
	680	612 – 748	300 – 360			220	−5	
Heavy duty industry gear oil	68	61.2 – 74.8	37.1 – 44.4		95	8	−8	For heavy duty gear, the gear surface stress is more than 1,100MPa, such as the gear lubrication of metallurgical, rolling, underground mining mechanical
	100	90 – 110	52.4 – 63.0					
	150	135 – 165	75.9 – 91.2					
	220	198 – 242	108 – 129			200		
	320	288 – 352	151 – 182					
	460	414 – 506	210 – 252					
	680	612 – 748	300 – 360			220		

15.2 Oil Indicator

Table 15-2 Oil indicator (dipstick)

mm

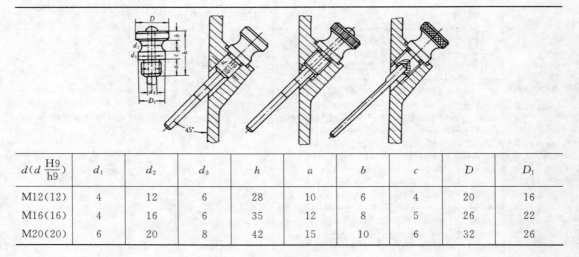

$d(d\frac{H9}{h9})$	d_1	d_2	d_3	h	a	b	c	D	D_1
M12(12)	4	12	6	28	10	6	4	20	16
M16(16)	4	16	6	35	12	8	5	26	22
M20(20)	6	20	8	42	15	10	6	32	26

15.3 Sealing

Table 15-3 Form and dimensions of felt oil seal

mm

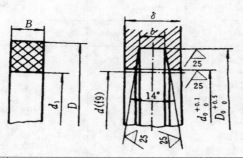

Marker example $d=50$ mm felt oil seal
Felt 50 JB/ZQ 4606—86

Shaft diameter d	Felt				Groove				
	D	d_1	B	Mass/kg	D_0	d_0	b	δ_{min}	
								For steel	For iron
15	29	14	6	0.001,0	28	16	5	10	12
20	33	19		0.001,2	32	21			

Chapter 15 Lubrication and Sealing

Continued

Shaft diameter d	Felt					Groove				
	D	d_1	B	Mass/kg		D_0	d_0	b	δ_{min}	
									For steel	For iron
25	39	24	7	0.001,8		38	26	6	12	15
30	45	29		0.002,3		44	31			
35	49	34		0.002,3		48	36			
40	53	39		0.002,6		52	41			
45	61	44		0.004,0		60	46			
50	69	49		0.005,4		68	51			
55	74	53		0.006,0		72	56			
60	80	58	8	0.006,9		78	61	7		
65	84	63		0.007,0		82	66			
70	90	68		0.007,9		82	66			
75	94	73		0.008,0		92	77			
80	102	78	9	0.011,0		100	82	8	15	18

Note: Felt oil seal suits to linear velocity $v<5$ m/s.

Table 15 – 4 Rotary-shaft lip seal

mm

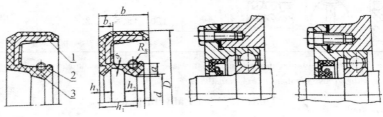

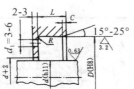

1 – frame
2 – compact hoop spring
3 – rubber seal body

B-type rotary-shaft lip seal without pair lip, shaft diameter $d=50$ mm
Basic diameter $D=72$ mm, basic width $b=8$ mm
Lip seal B 050072 GB 13781—92

Continued

Series of the basic size						Main size of the seal ring's section								
d	D	b	d	D	b	d	h	h_1	h_2	h_3	b_1	a	s	R_s
20	35,40,(45)	7	50	68,(70),72	8	20–30	6.2	5.1	2.8	0.2	1.9	2.6	0.8–1.2	0.8
22	35,40,47	7	55	72,(75),80	8	32–60	7.1	5.9	3.5	0.3	2.0	2.8	1.0–1.4	1.0
25	40,47,52	7	60	80,85	8	65–80	9.0	7.3	4.0	0.3	2.6	3.5	1.2–1.6	1.25
28	40,47,52	7	65	85,90	8	85–100	11.0	9.2	5.0	0.4	3.0	4.2	1.4–1.8	1.5
30	42,47,(50)	7	70	90,95	10	Size of the seal ring's groove								
30	52	7	75	95,100	10	b		$\leqslant 10$				>10		
32	45,47,52	7	80	100,110	10									
35	50,52,55	7	85	110,120	10	L		$>b+0.9$				$>b+1.2$		
38	55,58,62	8	90	(115),120	12									
40	55,(60),62	8	95	120	12	C		0.70–1.00				1.20–1.50		
42	55,62	8	100	125	12									
45	62,65	8				R		<0.50				<0.75		

Note: 1. B-type is main for single direction, FB-type is for bidirectional oil and dust sealing.
 2. Don't use value in brackets as possible.
 3. Seal ring section size is not standard size; disassembly hole number of d_1 is 3–4.

Table 15–5 Oil-ditch type seal groove

mm

Shaft diameter	R	t	b	d_1	a_{min}	h
25–80	1.5	4.5	4	$d_1=$ $d+1$	$a_{min}=$ $nt+R$	1
>80–120	2	6	5			
>120–180	2.5	7.5	6			
>180	3	9	7			

Note: 1. The size R, t, b in the table, in some individual case, can be used for non-corresponding shaft.
 2. Generally, the groove number is 2–4, mostly 3.

Table 15–6 Labyrinth seal

mm

d	10–50	>50–80	>80–110	>110~180
e	0.2	0.3	0.4	0.5
f	1	1.5	0.2	2.5

Chapter 16　Associated Component

16.1　Checking Hole and Checking Hole lid

Table 16-1　Checking hole and checking hole lid

mm

A	100,120,150,180,200
A_1	$A+(5-6)d_4$
A_2	$\frac{1}{2}(A+A_1)$
B	$B_1-(5-6)d_4$
B_1	Width of the hosing $-(15-12)$
B_2	$\frac{1}{2}(B+B_1)$
d_4	M6-M8, screw number is 4-6
R	5-10
h	3-5

Note: Material Q235A steel plate or HT150.

16.2　Ventilator

Table 16-2　Simple ventilator

mm

d	D	D_1	S	b	l	a	d_1
M12×1.25	18	16.5	14	19	10	2	4
M16×1.5	22	19.6	17	23	12	2	5
M20×1.5	30	25.4	22	28	15	4	6
M22×1.5	32	25.4	22	29	15	4	7
M27×1.5	38	31.2	27	34	18	4	8

Note: Material Q235A.

Table 16-3 Ventilator with filter screen

mm

d	d_1	d_2	d_3	d_4	D	h	a	b
M18×1.5	M33×1.5	8	3	16	40	40	12	7
M27×1.5	M48×1.5	12	4.5	24	60	54	15	10
d	c	h_1	R	D_1	S	K	e	f
M18×1.5	16	18	40	25.4	22	6	2	2
M27×1.5	22	24	60	39.6	32	7	2	2

Note: S—Wrench opening width.

16.3 Bearing Cover

Table 16-4 Screw-joined bearing cover

mm

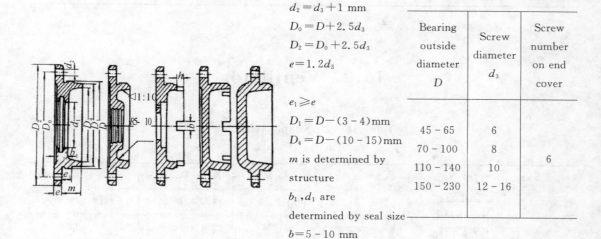

$d_2 = d_3 + 1$ mm
$D_0 = D + 2.5 d_3$
$D_2 = D_0 + 2.5 d_3$
$e = 1.2 d_3$

$e_1 \geqslant e$
$D_1 = D - (3 - 4)$ mm
$D_4 = D - (10 - 15)$ mm
m is determined by structure
b_1, d_1 are determined by seal size
$b = 5 - 10$ mm
$h = (0.8 - 1)b$

	d_3—connection screw diameter		
	Bearing outside diameter D	Screw diameter d_3	Screw number on end cover
	45 – 65	6	6
	70 – 100	8	
	110 – 140	10	
	150 – 230	12 – 16	

Note: Material HT150.

Chapter 16 Associated Component

Table 16-5 Embedded bearing cap

mm

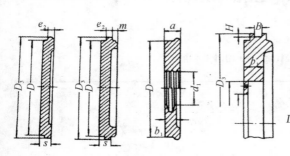

$e_2 = 5 - 10$ mm

$s = 10 - 15$ mm

m is determined by structure

$D_3 = D + e_2$, for O-type ring, it should be rounded according to O-type ring outside diameter.

D_5, d_1, b_1 and so on are determined is by seal size.

H, B determined by O-type ring groove

a is determined by structure

Note: Material HT150 or Q235A.

Table 16-6 Plug screw and oil sealing washer

mm

D	M14×1.5	M16×1.5	M20×1.5
D_0	22	26	30
L	22	23	28
l	12	12	15
a	3	3	4
D	19.6	19.6	25.4
S	17	17	22
D_1	≈0.95S		
d_1	15	17	22
H	2		

Note: Material of oil sealing washer: paronite, industrial development leather; material of plug screw: Q235A.

16.4 Oil Baffle

Table 16-7 Oil baffle (retainer)

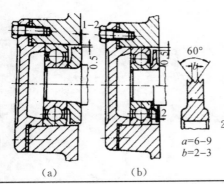

(a)   (b)

$a = 6-9$
$b = 2-3$

1. Scheme (a) is used for preventing grease in the bearing from being diluted and lost by lubricating oil in the housing. Its sealing effect is better. Scheme (a) is processed by turning, material is Q235A.

2. Scheme (b) is used for preventing oil from being extruded and entering the bearing when gears are meshing. Scheme (b) is stamped by Q235A steel plate.

16.5 Lifting Equipment

Table 16-8 Lifting eye and hook

(a) Lifting eye on casing cover

$d = b = (1.5 - 2.5)\delta_1$
$R = (1.0 - 1.2)d$
$e = (0.8 - 1)d$

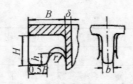

(b) Hook on lower housing

$b = (1.8 - 2.5)\delta$
$H = 0.8b$
$r = 0.25B$
$B = c_1 + c_2$
c_1, c_2 as in Table 5-1

Chapter 17 Structure of Cylindrical Gear

Table 17-1 Structure of cylindrical gear

No.	Structure form	Structure dimensions
1	Gear shaft	When $d_a < 2d$ or $e \leqslant 2.5 m_n$, the gear should be made into gear shaft
2	Solid gear	When $d_a \leqslant 200$ mm, the gear could be made into solid gear, which is made of rolling round steel or forged steel
3	Web gear	When $d_a < 500$ mm, the gear could be made into web gear, which is made of cast steel or forged steel. $D_3 = 1.6 d_s$ (steel) $D_1 = (D_0 + D_3)/2$ $D_3 = 1.7 d_s$ (iron) $D_0 = d_a - (10-14) m_n, n_1 = 0.5 m_n$ $D_2 = (0.25 - 0.35)(D_0 - D_3)$ (casting) $r \approx 0.5 C, C = (0.2 - 0.3) b$ $D_2 = 15 - 25$ mm (forging)

Continued

No.	Structure form	Structure dimensions
4	Spoke type gear	When $d_a > 400 - 1,000$ mm, spoke type should be used. $D_2 = d_a - 10 m_n$, $D_1 = 1.6 d_s$ (cast steel) $D_1 = 1.8 d_s$ (cast iron), $h = 0.8 d_n$, $h_1 = 0.8 h$, $\delta = 0.2 d_s$ $S = h/6$, $C = h/5$, $L = (1.2 - 1.5) d_s$, $n = 0.5 m_n$, $r = 0.5 C$
5	Combined gear	When the size of gear is very large, combined structure can be used, we can assemble rolling or forged rim to cast or forging steel wheel core by interference fit. To ensure connection reliable, add set screws in connecting zone. Wheel center structure sizes are the same with spoke type gear structure, rim sizes are as follows. $\delta_1 = 5$ mn, $D_2 = d_a - 18 m_n$, $\delta_2 = 0.2 \delta_1$ $l_1 = 0.28 \delta_1$, $d_3 = 0.05 d_s$, $l_3 = 0.05 d_s$ d_s—diameter of the shaft
6	Weld gear	When $d_a > 400$ mm, we can use weld gear. $D_1 = 1.6 d_s$, $l = (1.2 - 1.5) d_s$, $l \geq b$ $\delta_0 = 2.7$ mn, but it must be larger than 8 mm. $x = 5$ mm, $n = 0.5 m_n$, $C = (0.1 - 0.5) b$, but it must be larger than 8 mm. $S = 0.8 C$, $n = 0.5 m_n$, $D_0 = 0.5 (D_1 + D_2)$, $d_0 = 0.2 (D_2 - D_1)$ $K_a = 0.1 d_s$, $K_b = 0.5 d_s$, but it must be larger than 4 mm

Chapter 18 Limit and Fit, Tolerance and Surface Roughness

18.1 Limit and Fit

Table 18 - 1 Code of standard tolerance and fundamental deviation

Name		Code
Standard tolerance		IT1, IT2,..., IT18 levels totally
Basic deviation	Hole	A, B, C, CD, D, E, EF, F, FG, G, H, J, JS, K, M, N, P, R, S, T, U, V, X, Y, Z, ZA, ZB, ZC
	Shaft	a, b, c, cd, d, e, ef, fg, g, h, j, js, k, m, n, p, r, s, t, u, v, x, y, z, za, zb, zc

Table 18 - 2 Fitting type and code

Type	Hole based system H	Shaft based system h	Explanation
Clearance fit	a, b, c, cd, d, e, f, fg, g, h,	A, B, C, CD, D, E, F, FG, G, H	Gap decreased by order
Transition fit	j, js, k, m, n	J, JS, K, M, N	Compact by order
Interference fit	p, r, s, t, u, v, x, y, z, za, zb, zc	P, R, S, T, U, V, X, Y, Z, ZA, ZB, ZC	Compact by order

18.2 Standard Tolerance and Limit Deviation

Table 18-3 Standard tolerance value of basic size up to 500 mm

μm

Basic size/mm	Level							
	IT5	IT6	IT7	IT8	IT9	IT10	IT11	IT12
≤3	4	6	10	14	25	40	60	100
>3-6	5	8	12	18	30	48	75	120
>6-10	6	9	15	22	36	58	90	150
>10-18	8	11	18	27	43	70	110	180
>18-30	9	13	21	33	52	84	130	210
>30-50	11	16	25	39	62	100	160	250
>50-80	13	19	30	46	74	120	190	300
>80-120	15	22	35	54	87	140	220	350
>120-180	18	25	40	63	100	160	250	400
>180-250	20	29	46	72	115	185	290	460
>250-315	23	32	52	81	130	210	320	520
>315-400	25	36	57	89	140	230	360	570
>400-500	27	40	63	97	155	250	400	630

Table 18-4 Limit deviation values of holes of basic size from 10 mm to 315 mm

μm

Tolerance zone	Level	Basic size/mm							
		>10-18	>18-30	>30-50	>50-80	>80-120	>120-180	>180-250	>250-315
D	7	+68 +50	+86 +65	+105 +80	+130 +100	+166 +120	+185 +145	+216 +170	+242 +190
	8	+77 +50	+98 +65	+119 +80	+146 +100	+174 +120	+208 +145	+242 +170	+271 +190
	9	+93 +50	+117 +65	+142 +80	+174 +100	+207 +120	+245 +145	+285 +170	+320 +190
	10	+120 +50	+149 +65	+180 +80	+220 +100	+260 +120	+305 +145	+355 +170	+400 +190
	11	+160 +50	+195 +65	+240 +80	+290 +100	+340 +120	+395 +145	+460 +170	+510 +190

Chapter 18　Limit and Fit, Tolerance and Surface Roughness

Continued

Tolerance zone	Level	Basic size/mm							
		>10−18	>18−30	>30−50	>50−80	>80−120	>120−180	>180−250	>250−315
E	6	+43 +32	+53 +40	+66 +50	+79 +60	+94 +72	+110 +85	+129 +100	+142 +110
	7	+50 +32	+61 +40	+75 +50	+90 +60	+107 +72	+125 +85	+146 +100	+162 +110
	8	+59 +32	+73 +40	+89 +50	+106 +60	+126 +72	+148 +85	+172 +100	+191 +110
	9	+75 +32	+92 +40	+112 +50	+134 +60	+159 +72	+185 +85	+215 +100	+240 +110
	10	+102 +32	+124 +40	+150 +50	+180 +60	+212 +72	+245 +85	+285 +100	+320 +110
F	6	+27 +16	+33 +20	+41 +25	+49 +30	+58 +36	+68 +43	+79 +50	+88 +56
	7	+34 +16	+41 +20	+50 +25	+60 +30	+71 +36	+83 +43	+96 +50	+108 +56
	8	+43 +16	+53 +20	+64 +25	+76 +30	+90 +36	+106 +43	+122 +50	+137 +56
	9	+59 +16	+72 +20	+87 +25	+104 +30	+123 +36	+143 +43	+165 +50	+186 +56
H	5	+8 0	+9 0	+11 0	+13 0	+15 0	+18 0	+20 0	+23 0
	6	+11 0	+13 0	+16 0	+19 0	+22 0	+25 0	+29 0	+32 0
	7	+18 0	+21 0	+25 0	+30 0	+35 0	+40 0	+46 0	+52 0
	8	+27 0	+33 0	+39 0	+46 0	+54 0	+63 0	+72 0	+81 0
	9	+43 0	+52 0	+62 0	+74 0	+87 0	+100 0	+115 0	+130 0
	10	+70 0	+84 0	+100 0	+120 0	+140 0	+160 0	+185 0	+210 0
	11	+110 0	+130 0	+160 0	+190 0	+220 0	+250 0	+290 0	+320 0

Continued

Tolerance zone	Level	Basic size/mm							
		>10-18	>18-30	>30-50	>50-80	>80-120	>120-180	>180-250	>250-315
JS	6	±5.5	±6.5	±8	±9.5	±11	±12.5	±14.5	±16
	7	±9	±10	±12	±15	±17	±21	±23	26
	8	±13	±16	±19	±23	±27	±31	±36	±40
	0	±21	±26	±31	±37	±43	±50	±57	±65
N	7	−5 −23	−7 −28	−8 −33	−9 −10	−10 −45	−12 −52	−14 −60	−14 −66
	8	−3 −30	−3 −36	−3 −42	−4 −50	−4 −58	−4 −67	−5 −77	−5 −86
	9	0 −43	0 −52	0 −62	0 −74	0 −87	0 −100	0 −115	0 −130
	10	0 −70	0 −84	0 −100	0 −120	0 −140	0 −160	0 −185	0 −210
	11	0 −110	0 −130	0 −160	0 −190	0 −220	0 −250	0 −290	0 −320

Table 18 − 5　Limit deviation values of shafts of basic size from 10 mm to 315 mm

μm

Tolerance zone	Level	Basic size/mm													
		>10-18	>18-30	>30-50	>50-65	>65-80	>80-100	>100-120	>120-140	>140-160	>160-180	>180-200	>200-225	>225-250	>250-280	>280-315
d	6	−50 −66	−65 −78	−80 −96	−100 −119		−120 −142		−145 −170			−170 −199			−190 222	
	7	−50 −68	−65 −86	−80 −105	−100 −130		−120 −155		−145 −185			−170 −216			−190 −242	
	8	−50 −77	−65 −98	−80 −119	−100 −146		−120 −174		−145 −208			−170 −242			−190 −271	
	9	−50 −93	−65 −117	−80 −142	−100 −174		−120 −207		−145 −245			−170 −285			−190 −320	
	10	−50 −120	−65 −149	−80 −180	−100 −220		−120 −260		−145 −304			−170 −355			−190 −440	
	11	−50 −160	−65 −195	−80 −240	−100 −290		−120 −340		−145 −395			−170 −460			−190 −510	

Chapter 18 Limit and Fit, Tolerance and Surface Roughness

Continued

Tolerance zone	Level	Basic size/mm												
		>10 -18	>18 -30	>30 -50	>50 -65	>65 -80	>80 -100	>100 -120	>120 -140	>140 -160	>160 -180	>180 -200	>200 -225	>225 -250	>250 -280	>280 -315
f	7	−16 −34	−20 −41	−25 −50	−30 −60		−36 −71		−43 −83			−50 −96			−56 −108	
	8	−16 −43	−20 −53	−25 −64	−30 −76		−36 −90		−73 −106			−50 −122			−56 −137	
	9	−16 −59	−20 −72	−25 −87	−30 −104		−36 −123		−43 −143			−50 −165			−56 −186	
g	5	−6 −14	−7 −16	−9 −20	−10 −23		−12 −27		−14 −32			−15 −35			−17 −40	
	6	−6 −17	−7 −20	−9 −25	−10 −29		−12 −34		−14 −39			−15 −44			−17 −49	
	7	−6 −24	−7 −28	−9 −34	−10 −40		−12 −47		−14 −54			−15 −61			−17 −69	
h	5	0 −8	0 −9	0 −11	0 −13		0 −15		0 −18			0 −20			0 −23	
	6	0 −11	0 −13	0 −16	0 −19		0 −22		0 −25			0 −29			0 −32	
	7	0 −18	0 −21	0 −25	0 −30		0 −35		0 −40			0 −46			0 −52	
	8	0 −27	0 −33	0 −39	0 −46		0 −54		0 −63			0 −72			0 −81	
	9	0 −43	0 −52	0 −62	0 −74		0 −87		0 −100			0 −115			0 −130	
	10	0 −70	0 −84	0 −100	0 −120		0 −140		0 −160			0 −185			0 −210	
	11	0 −110	0 −130	0 −160	0 −190		0 −220		0 −250			0 −290			0 −320	
js	5	±4	±4.5	±5.5	±6.5		±7.5		±9			±10			±11.5	
	6	±5.5	±6.5	±8	±9.5		±11		±12.5			±14.5			±16	
	7	±9	±10	±12	±15		±17		±20			±23			±26	

Continued

Tolerance zone	Level	>10 −18	>18 −30	>30 −50	>50 −65	>65 −80	>80 −100	>100 −120	>120 −140	>140 −160	>160 −180	>180 −200	>200 −225	>225 −250	>250 −280	>280 −315
k	5	+9 +1	+11 +2	+13 +2	+15 +2	+15 +2	+18 +3	+18 +3	+21 +3	+21 +3	+21 +3	+24 +4	+24 +4	+24 +4	+27 +4	+27 +4
k	6	+12 +1	+15 +2	+23 +2	+21 +2	+21 +2	+25 +3	+25 +3	+28 +3	+28 +3	+28 +3	+33 +4	+33 +4	+33 +4	+36 +4	+36 +4
k	7	+19 +1	+23 +2	+27 +2	+32 +2	+32 +2	+38 +3	+38 +3	+43 +3	+43 +3	+43 +3	+50 +4	+50 +4	+50 +4	+56 +4	+56 +4
m	5	+15 +7	+17 +8	+20 +9	+24 +11	+24 +11	+28 +13	+28 +13	+33 +15	+33 +15	+33 +15	+37 +17	+37 +17	+37 +17	+34 +20	+34 +20
m	6	+18 +8	+21 +8	+25 +9	+30 +11	+30 +11	+35 +13	+35 +13	+40 +14	+40 +14	+40 +14	+46 +17	+46 +17	+46 +17	+52 +20	+52 +20
m	7	+25 +7	+29 +8	+34 +9	+41 +11	+41 +11	+48 +13	+48 +13	+55 +15	+55 +15	+55 +15	+63 +17	+63 +17	+63 +17	+72 +20	+72 +20
n	5	+20 +12	+24 +15	+28 +17	+33 +20	+33 +20	+38 +23	+38 +23	+45 +27	+45 +27	+45 +27	+51 +31	+51 +31	+51 +31	+57 +34	+57 +34
n	6	+23 +12	+28 +15	+33 +17	+39 +20	+39 +20	+45 +23	+45 +23	+52 +27	+52 +27	+52 +27	+60 +31	+60 +31	+60 +31	+66 +34	+66 +34
n	7	+30 +12	+36 +15	+42 +17	+50 +20	+50 +20	+58 +23	+58 +23	+67 +27	+67 +27	+67 +27	+77 +31	+77 +31	+77 +31	+86 +34	+86 +34
p	5	+26 +18	+31 +22	+37 +26	+45 +32	+45 +32	+52 +37	+52 +37	+61 +34	+61 +34	+61 +34	+70 +50	+70 +50	+70 +50	+79 +56	+79 +56
p	6	+29 +18	+35 +22	+42 +26	+51 +32	+51 +32	+59 +37	+59 +37	+63 +43	+63 +43	+63 +43	+79 +50	+79 +50	+79 +50	+88 +56	+88 +56
p	7	+36 +18	+43 +22	+51 +26	+62 +32	+62 +32	+72 +37	+72 +37	+83 +43	+83 +43	+83 +43	+96 +50	+96 +50	+96 +50	+108 +56	+108 +56
r	5	+31 +23	+37 +28	+45 +34	+54 +41	+56 +43	+66 +51	+69 +54	+81 +63	+83 +65	+86 +68	+97 +77	+100 +80	+104 +84	+117 +94	+121 +98
r	6	+34 +23	+41 +28	+50 +34	+60 +41	+62 +43	+73 +51	+76 +54	+88 +63	+90 +65	+93 +68	+106 +77	+109 +80	+113 +84	+126 +94	+130 +98
r	7	+41 +23	+49 +28	+59 +34	+71 +41	+73 +43	+86 +51	+89 +54	+103 +63	+105 +65	+108 +68	+123 +77	+126 +80	+130 +84	+146 +94	+150 +98

Chapter 18 Limit and Fit, Tolerance and Surface Roughness

Table 18-6 Suggested fit of main components in reducer

Fit components	Recommended fit	Disassembling method
The fit between gear, worm gear, pulley, sprocket, or coupling and shaft under general condition	H7/r6, H7/n6	Take use of presser
The fit between small bevel gear, the gear often disassembled, pulley, sprocket, or coupling and shaft	H7/m6, H7/k6	Driven into by press or hand hammer
The fit between worm gear's rim and wheel core	Tyre type H7/js6 Bolt connection type H7/h6	Pushed into by heating rim or presser
The fit between sleeve, oil baffle, or splashing baffle and shaft	D11/k6, F9/k6, F9/m6, h8/h7, H8/h8	
The fit between bearing sleeve cup and housing hole	H7/js6, H7/h6	
The fit between bearing cover and housing hole (or sleeve cup hole)	H7/d11, H7/h8	Assemble or disassemble by hand
The fit between embedded bearing cover's flange and housing hole's groove	H11/h11	
Tolerance zone of shaft segment contacting with seal	f9, h11	

Note: The fit between rolling bearing and shaft or hole refers to Table 13-4.

18.3 Surface Roughness

Table 18-7 Surface roughness and corresponding processing method

Roughness	∇	$R_a=25$	$R_a=12.5$	$R_a=6.3$	$R_a=3.2$	$R_a=1.6$	$R_a=0.8$	$R_a=0.4$	$R_a=0.2$
Surface state	Burrs are removed	Reamer tool marks can be seen a little	Processing trace can be seen	Processing trace can be seen a little	Processing trace can't be seen	The direction of processing trace can be seen	The direction of processing trace can be seen a little	The direction of processing trace can't be seen	Dark surface

Continued

Roughness	▽	$R_a=25$	$R_a=12.5$	$R_a=6.3$	$R_a=3.2$	$R_a=16$	$R_a=0.8$	$R_a=0.4$	$R_a=0.2$
Processing method	Cast, forge, stamping, hot rolling, cold rolling, power metallurgical	Rough turning, planing, vertical milling, flat milling, drill	Turning, boring, planing, drilling, flat milling, vertical milling, file, crude hinge, milling, gear milling	Turning, boring, planing, milling, shaving 1-2 polt/cm², tensile, milling, filing, rolling, milling, gear milling	Turning, boring, planing, milling, hinge, tensile, milling, rolling, gear milling, shaving 1-2 polt/cm²	Turning, boring, tensile, grinding, vertical milling, hinge, rolling, shaving 3-10 polt/cm²	Hinge, grinding, boring, tensile, rolling, shaving 3-10 polt/cm²	Cloth wheel grinding, grinding, super processing	Super processing

Table 18-8 Surface roughness of typical component

Surface characteristics	Position	Surface roughness R_a doesn't more than/μm		
Key and key groove	Working surface	6.3		
	Non-working surface	12.5		
Gear		Precision grade of the gear		
		7	8	9
	Tooth surface	0.8	1.6	3.2
	Outside circle	1.6-3.2		3.2-6.3
	End surface	0.8-3.2		3.2-6.3

Chapter 18 Limit and Fit, Tolerance and Surface Roughness

Continued

<table>
<tr><th colspan="2">Surface characteristics</th><th>Position</th><th colspan="3">Surface roughness R_a doesn't more than/μm</th></tr>
<tr><td rowspan="5">Mating surface of rolling bearing</td><td rowspan="3">Bearing seat hole diameter/mm</td><td colspan="4">Diameter tolerance grade of the shaft's and outside hole's fitting surface</td></tr>
<tr><td>IT5</td><td>IT6</td><td colspan="2">IT7</td></tr>
<tr><td>≤80
>80 – 500</td><td>0.4 – 0.8
0.8 – 1.6</td><td>0.8 – 1.6
1.6 – 3.2</td><td colspan="2">1.6 – 3.2
1.6 – 3.2</td></tr>
<tr><td colspan="2">End face</td><td>1.6 – 3.2</td><td colspan="2">3.2 – 6.3</td></tr>
<tr><td colspan="5"></td></tr>
<tr><td rowspan="2">Mating surface of shaft and hub such as driving components, coupling and so on</td><td colspan="2">Shaft</td><td colspan="3" rowspan="2">1.6 – 3.2</td></tr>
<tr><td colspan="2">Hub</td></tr>
<tr><td colspan="2">Non-mating surface of shaft end, chamfer, bolt hole and so on</td><td colspan="4">12.5 – 25</td></tr>
<tr><td rowspan="3">Surface of sealing</td><td colspan="2">Felt style</td><td>Rubber sealing type(lip type)</td><td>Oil ditch and labyrinth</td></tr>
<tr><td colspan="3">Peripheral velocity of the position connecting with shaft</td><td rowspan="2">1.6 – 3.2</td></tr>
<tr><td>≤3
0.8 – 1.6</td><td>>3 – 5
0.4 – 0.8</td><td>>5 – 10
0.2 – 0.4</td></tr>
<tr><td colspan="2">Subdivision surface of housing</td><td colspan="4">1.6 – 3.2</td></tr>
<tr><td colspan="2">Interface of checking hole and lid cover, bottom of housing</td><td colspan="4">6.3 – 12.5</td></tr>
<tr><td colspan="2">Location hole pin</td><td colspan="4">0.8 – 1.6</td></tr>
</table>

18.4 Accuracy Degree of Gear

Table 18-9 Values of F_r, F_w, f_f, f_{pt}, f_{pb} and F_β of the gear

μm

Pitch reference Diameter /mm		Normal module m_n/m	I tolerance group						II tolerance group									III tolerance group				
			Gear ring radial run-out tolerance F_r			Public normal length changing tolerance F_w			Tooth form tolerance f_f			Transverse pitch limit deviation $\pm f_{pt}$			Base pitch limit deviation $\pm f_{pb}$			Gear direction tolerance F_β				
From	To		Grade level															Gear width /mm		Level grade		
			6	7	8	6	7	8	6	7	8	6	7	8	6	7	8			6	7	8
—	125	≥1-3.5	25	36	45				8	11	14	10	14	20	9	13	18	—	40	9	11	18
		>3.5-6.5	28	40	50	20	28	40	10	14	20	13	18	25	11	16	22					
		>6.3-10	32	45	56				12	17	22	14	20	28	13	18	25	40	100	12	16	25
125	400	≥1-3.5	36	50	63				9	13	18	11	16	22	10	14	20					
		>3.5-6.5	40	56	71	25	36	50	11	16	22	14	20	28	13	18	25	100	160	16	20	32
		>6.3-10	45	63	86				13	19	28	16	22	32	14	20	30					

Chapter 19 Electromotor

Table 19 – 1 Technical data of Y series three-phase asynchronous motor

Type	Rated power kW	Rated current A	Speed r·min^{-1}	Efficiency %	Power factor (cosθ)	Locked rotor torque / Rated torque	Locked rotor current / Rated current	Maximum torque / Rated torque
Synchronous speed 3,000 r/min (grade 2)								
Y801-2	0.75	1.8	2,825	75	0.84	2.2	7.0	2.2
Y802-2	1.1	2.0	2,825	77	0.86	2.2	7.0	2.2
Y90S-2	1.5	3.4	2,840	78	0.85	2.2	7.0	2.2
Y90L-2	2.2	4.7	2,840	82	0.86	2.2	7.0	2.2
Y100L-2	3	6.4	2,880	82	0.87	2.2	7.0	2.2
Y112M-2	4	8.2	2,890	85.5	0.87	2.2	7.0	2.2
Y132S1-2	5.5	11.1	2,900	85.5	0.88	2.0	7.0	2.2
Y132S2-2	7.5	15	2,900	86.2	0.88	2.0	7.0	2.2
Y160M1-2	11	21.8	2,930	87.2	0.88	2.0	7.0	2.2
Y160M2-2	15	29.4	2,930	88.2	0.88	2.0	7.0	2.2
Y160L-2	18.5	35.5	2,930	89	0.89	2.0	7.0	2.2
Y180L-2	22	42.2	2,940	89	0.89	2.0	7.0	2.2
Synchronous speed 1,500 r/min (grade 4)								
Y801-4	0.55	1.5	1,390	73	0.76	2.2	6.5	2.2
Y802-4	0.75	2	1,390	74.5	0.76	2.2	6.5	2.2
Y90S-4	1.1	2.7	1,400	78	0.78	2.2	6.5	2.2
Y90L-4	1.5	3.7	1,400	79	0.79	2.2	6.5	2.2
Y100L1-4	2.2	5	1,420	81	0.82	2.2	7.0	2.2
Y100L2-4	3	6.8	1,420	82.5	0.81	2.2	7.0	2.2
Y112M-4	4	8.8	1,440	84.5	0.82	2.2	7.0	2.2
Y132S-4	5.5	11.6	1,440	85.5	0.84	2.2	7.0	2.2
Y132M-4	7.5	15.4	1,440	87	0.85	2.2	7.0	2.2
Y160M-4	11	22.6	1,460	88	0.84	2.2	7.0	2.2
Y160L-4	15	30.3	1,460	88.5	0.85	2.2	7.0	2.2
Y180M-4	18.5	35.9	1,470	91	0.86	2.0	7.0	2.2
Y180L-4	22	42.5	1,470	91.5	0.86	2.0	7.0	2.2

Continued

Type	Rated power kW	Rated current A	Speed r·min⁻¹	Efficiency %	Power factor (cosθ)	Locked rotor torque / Rated torque	Locked rotor current / Rated current	Maximum torque / Rated torque
\multicolumn{9}{c}{Synchronous speed 1,000 r/min (grade 6)}							
Y90S-4	0.75	2.3	910	72.5	72.5	0.70	2.0	2.0
Y90L-6	1.1	3.2	910	73.5	73.5	0.72	2.0	2.0
Y100L-6	1.5	4	940	77.5	77.5	0.74	2.0	2.0
Y112M-6	2.2	5.6	940	80.5	80.5	0.74	2.0	2.0
Y132S-6	3	7.2	960	83	83	0.76	2.0	2.0
Y132M1-6	4	9.4	960	84	84	0.77	2.0	2.0
Y132M2-6	5.5	12.6	960	85.3	85.3	0.78	2.0	2.0
Y160M-6	7.5	17	970	86	86	0.78	2.0	2.0
Y160L-6	11	24.6	970	87	87	0.78	2.0	2.0
Y180L-6	15	31.4	970	89.5	89.5	0.81	1.8	2.0
Y200L1-6	18.5	37.7	970	89.8	89.8	0.63	1.8	2.0
Y200L2-6	22	44.6	970	90.2	90.2	0.83	1.8	2.0
\multicolumn{9}{c}{Synchronous speed 750 r/min (grade 8)}							
Y132S-8	2.2	5.8	710	81	0.71	2.0	5.5	2.0
Y132M-8	3	7.7	710	82	0.72	2.0	5.5	2.0
Y132M1-8	4	9.9	720	84	0.73	2.0	6.0	2.0
Y160M2-8	5.5	13.3	720	85	0.74	2.0	6.0	2.0
Y160L-8	7.5	17.7	720	86	0.75	2.0	5.5	2.0
Y180L-8	11	25.1	730	86.5	0.77	1.7	6.0	2.0
Y200L-8	15	34.1	730	88	0.76	1.8	6.0	2.0
Y225S-8	18.5	41.3	730	88	0.76	1.8	6.0	2.0
Y225M-8	22	47.6	730	90	0.78	1.8	6.0	2.0

Note: S—short frame, M—middle frame, L—long frame.

Table 19-2 Shape and installation size of Y series three-phase asynchronous motor

Frame cushion number	D (DA)	E (EA)	F (FA)	GD (GF)	G (GB)	A	AA	AB	AC	AD	B	BB	C	CA	H	HA	HC	HD	K	L 2P	L 4,6,8,10	LC 2P	LC 4,6,8,10P	LD
80	19j6	40	6	6	15.5	125	37	165	165	150	100	135	50	100	$80_{-0.5}^{0}$	13	170	—	10	235	235	332	332	
90S	24j6	50	8	7	20	140	37	180	175	155	100	135	56	110	$90_{-0.5}^{0}$	13	190	—	10	310	310	368	368	
90L	24j6	50	8	7	20	140	37	180	175	155	125	160	56	110	$90_{-0.5}^{0}$	13	190	—	10	335	335	393	393	
100L	28j6	60	8	7	24	160	42	205	205	180	140	180	63	120	$100_{-0.5}^{0}$	15		245	12	380	380	445	445	
112M	28j6	60	8	7	24	190	52	245	230	190	140	185	70	131	$112_{-0.5}^{0}$	17		265	12	400	400	463	463	
132S	38k6	80	10	8	33	216	63	280	270	210	178	205	89	168	$132_{-0.5}^{0}$	20		315	12	475	475	559	559	
132M	38k6	80	10	8	33	216	63	280	270	210	210	213	89	168	$132_{-0.5}^{0}$	20		315	12	515	515	597	597	
160M	42k6	110	12	8	37	254	73	330	325	255	210	275	108	177	$160_{-0.5}^{0}$	22	—	385	15	600	600	717	717	55
160L	42k6	110	12	8	37	254	73	330	325	255	254	320	108	177	$160_{-0.5}^{0}$	22	—	385	15	645	645	761	761	55
180M	48k6	110	14	9	42.5	279	73	355	360	285	241	315	121	199	$180_{-0.5}^{0}$	24		420	15	670	670	783	783	86
180L	48k6	110	14	9	42.5	279	73	355	360	285	279	353	121	199	$180_{-0.5}^{0}$	24		420	15	710	710	821	821	105
200L	55m6	110	16*	10*	49*	318	73	395	400	310	305	378	133	221	$200_{-0.5}^{0}$	27		475	19	75	75	81	81	102

Note: 1.The distance between the second extension shoulder and air hood is about 8mm,the shape size of L, LC is maximum.

2.The size of * is the first shaft extension's and the second shaft extension's size of 4,6,8P's motor; when frame number is 200L and motor is 2P,the second shaft extension's FA=14 mm, GF=9 mm,GB=42.5 mm

Chapter 20　Drawing Examples and Cases

20.1　Drawing Examples

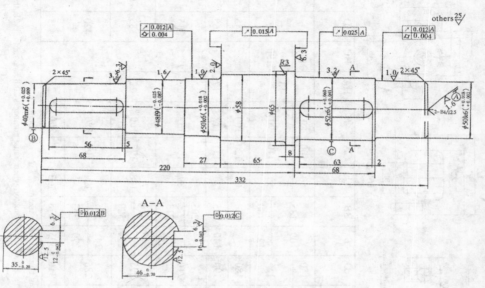

Technical conditions:
1. Quenched and tempered 217 – 255 HBS.
2. Other fillet $R = 1.5$ mm.

Mark	Positions	Partition	Modified number	Signature	Date					(Institution name)
						45				
Design		Date	Standardization			Stage mark		Quality	Scale	Shaft
Trace									1 : 1	
Review										(Pattern code)
Process			Approval			Total				

Fig. 20 - 1　Shaft drawing

Chapter 20 Drawing Examples and Cases

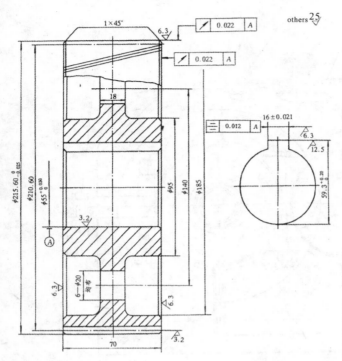

Normal modulus	m_n	2.5
Number of teeth	z_1	81
Tooth angle	α	20°
Tooth high coefficient	h_a	1.0
Helix angle	β	15° 56′ 33″
Helix direction		left
Displacement coefficient	x	0
Accuracy level		8HJGB10095.1988
Center distance	$a \pm f_a$	130±0.031
Matching gears	Drawing number	
	Number of teeth	z_2	19
Tolerance group	Check item	Tolerance value
I	F_r	0.063
	F_w	0.050
II	F_{pt}	±0.022
	f_i	0.018
III	F	0.025
Tooth thickness	The average length of the common law and its upper and lower deviation	$80.752_{-0.220}^{-0.176}$
	Spanned-tooth number	K	11

Technical conditions:
1. Normalizing treatment 162 – 217 HBS.
2. Other chamfer 2×45°, fillet $R=5$ mm.

Mark	Positions	Partition	Modified number	Signature	Date					(Institution name)
Design		Date	Standardization			Stage mark		Quality	Scale	Gear
Trace									1:1	
Review										(Pattern code)
Process			Approval			Total				

Fig. 20 – 2 Gear drawing

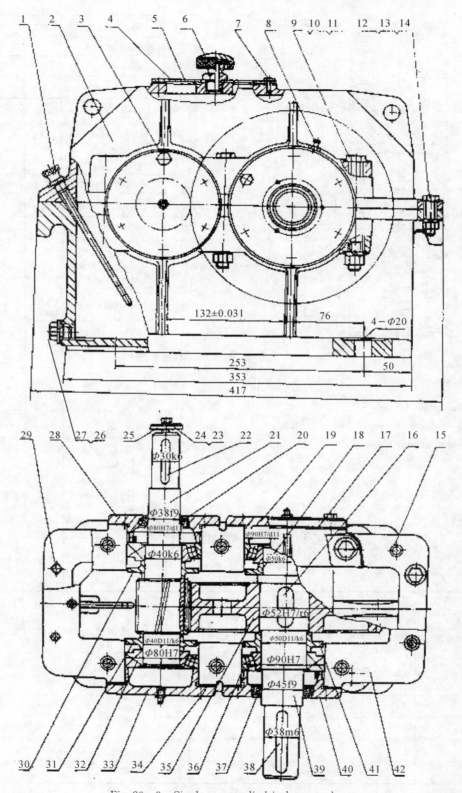

Fig. 20-3 Single stage cylindrical gear reducer

Chapter 20 Drawing Examples and Cases

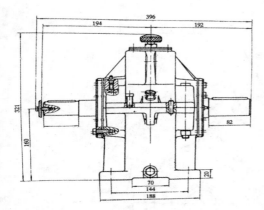

Technical feature

Input power/ kW	Shaft speed/ (r·min^{-1})	Total drive ratio	Efficiency	Transmission characteristics				
				β	m_n	Number of teeth		Accuracy level
2.795	384	4.021	0.95	9°59′12″	2.5	z_1	21	8GJ
						z_2	85	8HK

Technical requirements:

1. Before assembly, fit dimensions should be checked according to drawings. All components should be cleaned withkerosene and bearings should be cleaned by gasoline, there should not be any litter in the housing, and oil resistant paint should be painted on the inside wall of the housing.

2. Oil leakage should be avoided on the split surface, contacting surface and sealing surface. Sealant or water glass can be used on the split surface and other filler should not be used.

3. There should be an axial clearance of 0.05 – 0.1 mm for bearing.

4. After assembly, contact pits should be checked by coloring method, along gear height not more than 30%, along gear length not more than 50%.

5. The oil in the housing is 150 middle-load industry oil, and the depth is rational.

6. Light grey paint is painted on the outside surface of the housing.

7. Experiments should be done according to the experiment procedure.

No.	Code	Name	Quantity	Material	Singleton Quality	Total	Remarks
42		Housing	1	HT200			
41		Oil baffle disc	1	Q235A			
40		Bearing cover	1	HT200			
39		shaft	1	45			
38	GB 1096—79	Key 10×70	1				
37	GB 1387—92	SealingringFB045065	1				
36	GB/T 297—94	30210	2				
35		Gear	1	45			
34		Adjusting shim	2	08F			
33		Bearing cover	1	HT200			
32	GB/T 297—94	30208	2				
31		Oil baffle	1	Q235A			
30		Oil baffle	1	Q235A			
29		Separating screw	1				
28		Bearing cover	1	HT200			
27		Shim	1	Asbestos rubber sheet			
26		Plug M16×1.5	1	Q235A			
25	GB 892—86	End cap B35	1	Q235A			
24	GB 892—87	Washer 6	1	65Mn			
23	GB 5783—86	Bolt M6×20	1	8.8 level			
22	GB 1096—79	Key 8×50	1	45			
21		Gear shaft	1	45			
20	GB 13871—92	Seal ringFB035055	1				
19		Bearing cover	1	HT200			
18		Oil baffle	1	Q235A			
17		Adjusting shim	2	08F			
16	GB 1096—79	Key 16×45	1	45			
15	GB 117—86	Pin 8×35	1	45			
14	GB 93—87	Washer 10	2	62Mn			
13	GB 6170—86	Nut M10	2	8level			
12	GB 5783—86	Bolt M10×40	2	8.8level			
11	GB 93—87	Washer 12	6	65Mn			
10	GB 6170—86	Nut M12	6	8level			
9	GB 5782—86	Bolt M12×120	4	8.8level			
8	JB/T 7940.1—95	Oil cup M8×1	4				
7	GB 5783—86	Bolt M6×12	4	8.8level			
6		VenlitatorM18×1.5	1				
5		Check hole lid	1	Q235A			
4		Shim	1	Soft board			
3	GB 5783—86	Bolt M8×25	24	8.8level			
2		Cover	1	HT200			
1		Oil gauge	1	Q235A			

Continued Fig. 20-3 Single stage cylindrical gear reducer

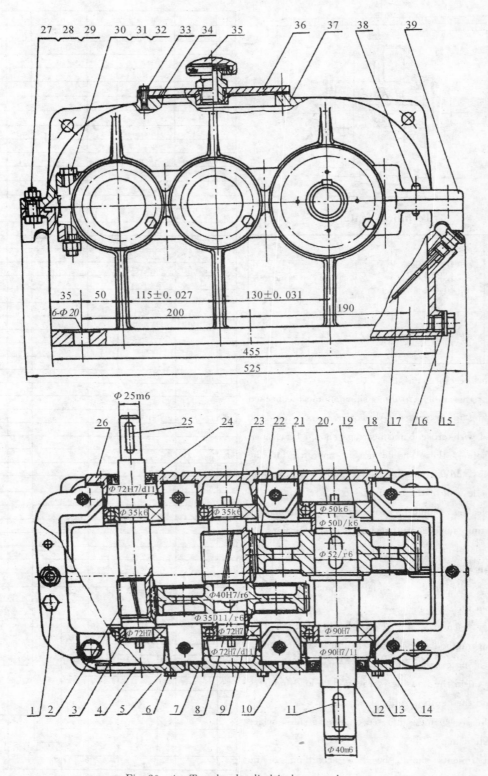

Fig. 20-4 Two level cylindrical gear reducer

Chapter 20 Drawing Examples and Cases

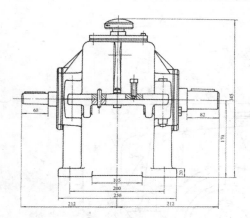

Technical feature

Input power/ kW	Shaft speed/ (r·min⁻¹)	Total drive ratio i	Efficiency	Transmission characteristics (two levels)			
				β	m_n	Number of teeth	Accuracy level
1.4	1,400	0.94	12.5	10°42′	2	z_1 23	8GB
						z_2 90	8HJ
				15°56′	2.5	z_1 19	8GB
						z_2 81	8HJ

Technical conditions:

1. Before the assembly, the unmachined surface of the housing and other casting parts should be cleaned, burrs should be removed, and antirust liquid should be used.

2. Before the assembly, components should be cleaned by kerosene, bearings by gasoline cleaning, after drying the fit surface should be oiled.

3. Oil leakage should be avoided on the split surface, contacting surface and sealing surface. sealant or water glass can be used on the split surface and other filler should not be used.

4. After assembly, contact pits should be checked by coloring method, along gear height not more than 30%, along gear length not more than 50%., the backlash: the first grade $j_{nmin}=0.140$ mm, second grade $j_{nmin}=0.160$ mm.

5. When adjust and fix bearings, there should be an axial clearance of 0.2 - 0.5 mm.

6. The gear reducer is equipped with 220 industrial gear oil, oil reaches the specified depth.

7. The inside wall should be painted with anti oil paint, reducer surface painted with grey paint.

8. Experiments should be done according to the experiment procedure.

Serial number	Code	Name	Number	Material Science	Single quality	Total quality	Remarks
39		Housing	1	HT150			
38	GB 117—86	Pin 8×35	2	35			
37		Housing cover	1	HT150			
36		Check hole cover	1	Q235A			
35		Ventilator	1				
34	QB 365—81	Shim	1	Mild steel board			
33	GB 5783—86	Bolt M6 * 20	6	8.8level			
32	GB 6170—86	Nut M12	8	8level			
31	GB 93—87	Washer 12	8	65Mn			
30	GB 5782—86	Bolt M12 * 10	8	8.8level			
29	GB 6170—86	Nut M10	2	8level			
28	GB 93—87	Washer 10	2	65Mn			
27	GB 5782—86	Bolt M10 * 35	2	8.8level			
26		Bearing cover	1	HT150			
25	GB 1096—79	Key 8 * 45	1	45			
24	GB 13871—92	Sealing ring FB032052	1				
23		Gear shaft	1	45			
22		Gear	1	45			
21	GB/T 276—94	Deep groove ball bearing 3210	2				
20	GB/T 1096—79	Key 16 * 36	1	45			
19		Sleeve	1	Q235A			
18		Bearing Cover	1	HT150			
17		M16 dipstick	1	Q235A			
16		Sealing shim	1	Asbestos rubber sheet			
15		Plug M20 * 1.5	1	Q235A			
14		Through cover	1	HT150			
13	GB 13871—92	Sealing ring B045065	1				
12		Shaft	1	45			
11	GB 1096—79	Key 12 * 50	1	45			
10		Shim	2	08F			
9		Gear	1	45			
8	GB 1096—79	Key 12 * 28	1	45			
7		Sleeve	1	45			
6	GB 5783—86	Bolt M8 * 20	36	8.8level			
5		Shim	4	08F			
4		Bearing cover	3	HT150			
3	GB/T 276—94	6207	4				
2		Gear shaft	1	45			
1	GB 85—88	Separating screw	1	8.8level			

Continued Fig. 20 - 4 Two level cylindrical gear reducer

Course Design for Machinery Design

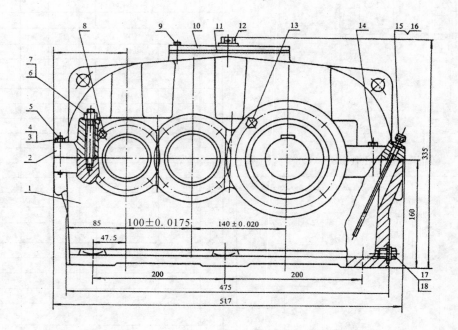

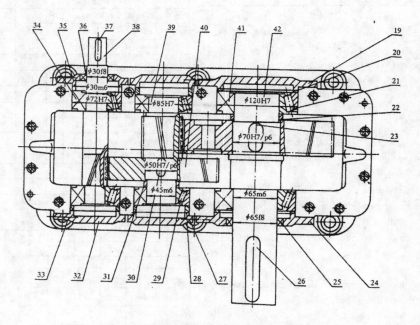

Fig. 20-5 Two stage cylindrical gear reducer

Chapter 20 Drawing Examples and Cases

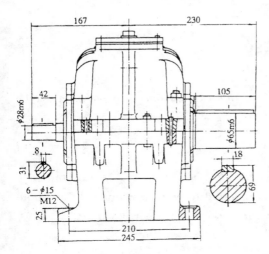

Technical feature

Input power/ kW	Shaft speed/ (r·min^{-1})	Total drive ratio i	Efficiency	Transmission characteristics			
				β	m_n	Number of teeth	Accuracy level
1.4	1,400	0.94	12.5	16°15′	2	z_1 21	6HK
						z_2 75	6JL
				15°21′	3	z_1 20	6HK
						z_2 70	6KM

Technical conditions:

1. The inner race of bearing should be close to shoulder, the feeler of 0.05 mm should not pass.

2. The axial clearance of tapered roller bearing: 0.04–0.07 mm for input shaft, 0.05–0.10 mm for middle shaft, 0.08–0.15mm for output shaft.

3. The minimum backlash for gear meshing: first level 0.140 mm, second level 0.160 mm.

4. After assembly, contact pits should be checked by coloring method, along gear height not more than 30%, along gear length not more than 50%.

5. The oil in the housing is 320 middle-load industry oil, and the depth is rational.

6. Anti loose glue should be used for fastening bolt, sealing glue for joined surface and paint for outside surface.

7. In idling test, the speed of high speed shaft is 750–1,500r/min, 2 hours each for forward and reverse, the operation should be stable and there is no loosening for connected fasteners and no oil leakage for sealing position.

8. Experiments should be done according to the experiment procedure.

No.	Code	Name	Number	Material
35		Bearing cover	1	HT200
34	GB/T 297—94	Tapered roller bearing 32306	2	
33		Adjusting ring	2	45
32		Bearing cover	1	HT200
31		Sleeve	1	45
30	GB 1096—79	Key 14 * 30	1	45
29	GB/T 297—94	32209	2	
28		Adjusting ring	2	45
27		Bearing cover	2	HT200
26	GB 1096—79	Key 18 * 100	1	45
25	GB 13871—92	Sealing ring	1	
24		Bearing cover	1	HT200
23	GB 1096—79	Key 20 * 45	1	45
22		Sleeve	1	45
21	GB 273—94	32213	2	
20		Adjusting ring	2	45
19		Bearing cover	1	HT200
18		Washer 20	1	Asbestos rubber sheet
17		Plug M20 * 1.5	1	Q235A
16	GB 3452.1—92	O type 9 * 1.8	1	
15		Dipstick	1	Q235A
14	GB 5783—86	Bolt M8 * 30	2	8.8level
13	GB 5783—86	Bolt M12 * 25	8	8.8level
12		Ventilator plug M20 * 1.5	1	Q235A
11		Shim	1	Oil resistant rubber asbestos board
10		Checking hole cover	1	Q235A
9	GB 782—86	Bolt M6 * 15	6	8.8level
8	GB 5782—86	Bolt M10 * 20	16	8.8level
7	GB 899—88	Stud AM10 * 75	8	8.8level
6	GB 6170—86	Nut M10	8	8level
5	GB 117—86	Pin 6 x 50	2	45
4	GB 6170—86	Nut M8	4	8level
3	GB 5782—86	Nut M8 * 50	4	8.8level
2		Cover	1	HT200
1		Housing	1	HT200

Continued Fig. 20-5 Two stage cylindrical gear reducer

20.2 Case 1

This case was done by a student, the design and calculations are not completely correct, but the design procedure can be referred to.

20.2.1 Design data

As shown in Fig. 20-6, force on belt $F = 2.8$ kN, belt velocity $v = 1.2$ m/s and diameter of the roller $D = 350$ mm. The reducer is required to work continuously, without reverse under stable load and clean environment, for 8 years (16 h per day).

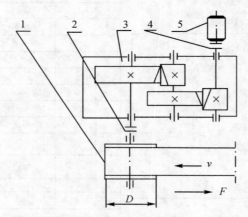

Fig. 20-6 Transmission of a belt conveyor

20.2.2 Design procedure and results

1. Selection and calculation of motor

Choose three phase asynchronous motor according to the working conditions and job requirements. Open structure, power 380 V, Y series.

The power of working part:

$$P_w = \frac{Fv}{1,000} = 2,800 \times 1.2/1,000 = 3.36 \text{ kW}$$

(1) Calculation of motor power.

The total transmission efficiency:

$$\eta = \eta_{gear}^2 \cdot \eta_{bearing}^4 \cdot \eta_{coupling}^2 \cdot \eta_{rollor}$$

Because there are 4 pairs of bearings, 2 pairs of couplings, 1 roller, and 2 pairs of gear, according to Table 10-1, choose:

Gear meshing efficiency:

$$\eta_{gear} = 0.97 \text{ (accuracy grade: 8)}$$

Chapter 20　Drawing Examples and Cases

Efficiency of bearing:
$$\eta_{bearing} = 0.98 \text{ (single-row tapered roller bearing)}$$

Efficiency of coupling:
$$\eta_{coupling} = 0.99$$

Efficiency of roller:
$$\eta_{roller} = 0.96$$

The total transmission efficiency:
$$\eta = 0.97^2 \times 0.98^4 \times 0.99^2 \times 0.96 = 0.816,6$$

The power required:
$$P_r = P_w/\eta = 3.36/0.816,6 = 4.114,6 \text{ kW}$$

According to Table 19-1, choose three phase asynchronous motor model: Y132S-4, rated power: $P_0 = 5.5$ kW, or choose Y132M2-6, rated power: $P_0 = 5.5$ kW.

(2) Determine the speed of working part.

Speed of the roller:
$$n_W = \frac{60v}{\pi D} = 60 \times 1.2/(3.14 \times 0.35) = 65.5 \text{ r/min}$$

According to the Table 19-1, the comparison of synchronous speed of 1,500 r/min and 1,000 r/min and the date and total transmission ratios are list in the following Table 20-1.

Table 20-1　Total transmission ratio and data of the motors

Plan No.	Type of motor	Rated power/kW	Synchronous speed /(r·min^{-1})	Speed with full load	Quality of motor	Total speed ratio
1	Y132S-4	5.5	1,500	1,440	51	21.98
2	Y132M2-6	5.5	1,000	960	73	14.66

Compare the two plans, although the motor of plan 1 is in low price with big transmission ratio, we decide to select plan 2 for its compact structure.

Its parameter:

Type: Y132M2-6.

Rated power: 5.5 kW.

Synchronous speed: 1,000 r/min.

The motor's center height is 132 mm and the extended shaft is $D \times E = 38$ mm $\times$ 80 mm according to Table 19-2.

2. Distribution of drive ratio

The total transmission ratio of reducer $i_{reducer} = 14.66$ according to Table 20-1.

Take the high-level gear transmission ratio:
$$i_1 = \sqrt{1.35 i_{reducer}} = \sqrt{1.35 \times 4.66} = 4.448,2$$

So the low-level gear transmission ratio:
$$i_2 = i_{reducer}/i_1 = 14.66/4.448,2 = 3.294,9$$

3. Selection and calculation of kinematic and dynamic parameters

(1) Parameters of the motor shaft (0 shaft).
$$P_0 = P_r = 4.114,6 \text{ kW}$$
$$n_0 = 960 \text{ r/min}$$
$$T_0 = \frac{9.55 P_0}{n_0} = 9.55 \times 4.114,6 \times 10^3 / 960 = 40.93 \text{ N} \cdot \text{m}$$

(2) Parameters of the high-speed shaft (I shaft).
$$P_1 = P_0 \eta_{01} = P_0 \eta_{\text{coupling}} = 4.114,6 \times 0.99 = 4.073 \text{ kW}$$
$$n_1 = n_0 = 960 \text{ r/min}$$
$$T_1 = \frac{9.55 P_1}{n_1} = 9.55 \times 4.073 \times 10^3 / 960 = 40.518 \text{ N} \cdot \text{m}$$

(3) Parameters of the middle shaft (II shaft).
$$P_2 = P_1 \eta_{12} = P_1 \eta_{\text{gear}} \eta_{\text{bearing}} = 4.073 \times 0.97 \times 0.98 = 3.872 \text{ kW}$$
$$n_2 = \frac{n_1}{i_{12}} = \frac{960}{4.448,2} = 215.82 \text{ r/min}$$
$$T_2 = \frac{9.55 P_2}{n_2} = 9.55 \times 3.872 \times 10^3 / 215.82 = 171.335 \text{ N} \cdot \text{m}$$

(4) Parameters of the low-speed shaft (III shaft).
$$P_3 = P_2 \eta_{23} = P_2 \eta_{\text{gear}} \eta_{\text{bearing}} = 3.872 \times 0.97 \times 0.98 = 3.681 \text{ kW}$$
$$n_3 = \frac{n_2}{i_{23}} = \frac{215.82}{3.294,9} = 65.5 \text{ r/min}$$
$$T_3 = \frac{9.55 P_3}{n_3} = 9.55 \times 3.681 \times 10^3 / 65.5 = 536.695 \text{ N} \cdot \text{m}$$

(5) Parameters of the roller shaft (IV shaft).
$$P_4 = P_3 \eta_{34} = P_3 \eta_{\text{coupling}} \eta_{\text{bearing}} = 3.681 \times 0.99 \times 0.98 = 3.571 \text{ kW}$$
$$n_4 = n_3 = 65.5 \text{ r/min}$$
$$T_4 = \frac{9.55 P_4}{n_4} = 9.55 \times 3.571 \times 10^3 / 65.5 = 520.66 \text{ N} \cdot \text{m}$$

The results summarized above are listed in Table 20-2.

Table 20-2 Kinematic and dynamic parameters of each shaft

Shaft No.	Power /kW	Rotating speed / (r · min^{-1})	Torque/ (N · m)	Drive type	Speed ratio	Efficiency
0	4.114,6	960	40.93	Coupling	1	0.99
I	4.073	960	40.518	Gear	4.448,2	0.97
II	3.872	215.82	171.335	Gear	3.294,9	0.97
III	3.681	65.5	536.695	Coupling	1	0.99
IV	3.571	65.5	520.66			

4. Design and calculation of transmission components

(1) Design and calculation of high-speed stage gear.

1) Basic parameters of gear.

(a) Small and big gears of high-speed stage.

Small gear of high-speed stage:

Material: 40Cr, tempering.

Rigidity: 280HBS.

Number of teeth: $z_1 = 22$.

Big gear of high-speed stage:

Material: 45 steel, tempering.

Rigidity: 240HBS.

Number of teeth: $z_2 = 4.448, 2 \times 22 = 97.860, 4$, so $z_2 = 98$.

(b) Precision grade of the gear.

Choose grade 8 according to reference [1].

2) Calculation of the contact strength of gear.

Design by the tooth surface contact strength, according to the formula:

$$d_{1t} \geqslant \sqrt[3]{\frac{2K_t T_1}{\varphi_d \varepsilon_a} \times \frac{u+1}{u} \times \left(\frac{Z_H Z_E}{[\sigma_H]}\right)^2}$$

(a) Determine the variables in the formula.

I. Choose load factor:

$$K_t = 1.6$$

II. Choose zone factor:

$$Z_H = 2.433$$

III. According to the required figure in reference [1]:

$$\varepsilon_{a1} = 0.765, \quad \varepsilon_{a2} = 0.88 \Rightarrow \varepsilon_a = 0.765 + 0.88 = 1.645$$

IV. Torque of the small gear:

$$T_1 = 9.55 P_1 / n_1 = 9.55 \times 4.073 \times 10^3 / 960 = 40.518 \text{ N} \cdot \text{m}$$

V. According to the required figure in reference [1], the tooth width factor:

$$\varphi_d = 1$$

VI. The elastic influence coefficient:

$$Z_E = 189.8 \text{ MPa}^{\frac{1}{2}}$$

VII. The contact strength limit of the small gear:

$$\sigma_{Hlim1} = 600 \text{ MPa}$$

The contact strength limit of the big gear:

$$\sigma_{Hlim2} = 550 \text{ MPa}$$

VIII. Calculate the number of stress cycles by the formula:

$$N_1 = 60n_1jL_h = 60 \times 960 \times 1 \times (2 \times 8 \times 300 \times 8) = 2.211,8 \times 10^9 \text{ h}$$
$$N_2 = 60n_2jL_h = 60 \times 215.82 \times 1 \times (2 \times 8 \times 300 \times 8) = 4.972,5 \times 10^8 \text{ h}$$

IX. Calculate the contact fatigue allowable stress.

Safety factor $S=1$. By the required formula:
$$[\sigma_H]_1 = \frac{K_{HN1}\sigma_{Hlim1}}{S} = 0.93 \times 550 = 511.5 \text{ MPa}$$
$$[\sigma_H]_2 = \frac{K_{HN2}\sigma_{Hlim2}}{S} = 0.96 \times 450 = 432 \text{ MPa}$$
$$\Rightarrow$$
$$[\sigma_H] = ([\sigma_H]_1 + [\sigma_H]_2)/2 = (511.5 + 432)/2 = 471.75 \text{ MPa}$$

(b) Calculate.

I. Calculate the pitch circle diameter of the small gear.
$$d_{1t} \geqslant \sqrt[3]{\frac{2K_tT_1}{\varphi_d\varepsilon_\alpha} \times \frac{u+1}{u} \times (\frac{Z_HZ_E}{[\sigma_H]})^2} =$$
$$\sqrt[3]{\frac{2 \times 1.6 \times 40.518 \times 10^3}{1 \times 1.645} \times \frac{4.448,2}{3.448,2} \times (\frac{2.433 \times 189.8}{471.75})^2} = 46.013 \text{ mm}$$

II. Calculate the circular velocity.
$$\frac{3.14 \times 46.013 \times 960}{60 \times 1,000} = 2.311,7 \text{ m/s}$$

III. Calculation of tooth width and module.
$$b = \varphi_d \times d_{1t} = 46.013 \text{ mm}, \quad m_{nt} = \frac{d_{1t}\cos\beta}{Z_1} = \frac{46.013 \times \cos 14°}{24} = 1.86 \text{ mm}$$
$$h = 2.25m_{nt} = 2.25 \times 1.86 = 4.185 \text{ mm}, \quad \frac{b}{h} = b/h = \frac{46.013}{4.185} = 10.995$$

IV. Calculation of overlap contact ratio:
$$\varepsilon_\beta = 0.318\varphi_d z_1\tan\beta = 0.318 \times 1 \times 22 \times \tan 14° = 1.744$$

V. Calculation of the load factor.

According to $v = 2.311,7$ m/s, accuracy of 8, we get the dynamic factor:

According to the table in reference [1], $K_{H\alpha} = K_{F\alpha} = 1.4, K_A = 1$.

According to the table in reference [1], we get $K_v = 1.07$.

$K_{H\beta} = 1.15 + 0.18\varphi_d^2 + 0.31 \times 10^{-3}b = 1.15 + 0.18 + 0.31 \times 10^{-3} \times 46.013 = 1.344$

According to the table in reference [1], we get: $K_{F\beta} = 1.28$.

So the load factor:
$$K = K_AK_vK_{H\alpha}K_{H\beta} = 1 \times 1.07 \times 1.4 \times 1.344 = 2.013,3$$

VI. Modify the pitch circle diameter by the actual load factor:
$$d_1 = d_{1t}\sqrt[3]{\frac{K}{K_t}} = 46.013 \times \sqrt[3]{\frac{2.013,3}{1.6}} = 49.675,6 \text{ mm}$$

VII. Calculate the module.

$$m_n = \frac{d_1 \cos\beta}{z_1} = \frac{49.675,6 \times \cos 14°}{22} = 2.191 \text{ mm}$$

3) Calculation of the bending strength of gear.

According to the formula:

$$m_n \geq \sqrt[3]{\frac{2KT_1 Y_\beta \cos^2\beta}{\varphi_d z^2_{\ 1} \varepsilon_a} \left(\frac{Y_{Fa} Y_{Sa}}{[\sigma_F]}\right)}$$

(a) Determine the parameters in the formula:

I. Calculate load factor.

$$K = K_A K_v K_{F\alpha} K_{F\beta} = 1 \times 1.07 \times 1.4 \times 1.28 = 1.917$$

II. According to the overlap contact ratio $\varepsilon_\beta = 1.744$, we get $Y_\beta = 0.75$.

III. Calculate equivalent number of teeth.

$$z_{v1} = z_1 / \cos^3\beta = 30/\cos^3 14° = 24.08$$
$$z_{v2} = z_2 / \cos^3\beta = 99/\cos^3 14° = 107.28$$

IV. Choose form factor from the table in reference [1].

$$Y_{Fa1} = 2.72, \quad Y_{Fa2} = 2.18$$

V. Choose stress correction factor from the table in reference [1].

$$Y_{Sa1} = 1.57, \quad Y_{Sa2} = 1.79$$

VI. Choose the bending fatigue limit from figure in reference [1].

$$\sigma_{FE1} = 500 \text{ MPa}, \quad \sigma_{FE2} = 380 \text{ MPa}$$

VII. Choose bending fatigue life factor from the figure in reference [1].

$$K_{FN1} = 0.86, \quad K_{FN2} = 0.93$$

VIII. Calculate the allowable bending stress fatigue, bending fatigue safety factor $S = 1.4$,

$$[\sigma_F]_1 = \frac{K_{FN1} \sigma_{FF1}}{S} = \frac{0.86 \times 500}{1.4} = 307.14 \text{ MPa}$$

$$[\sigma_F]_2 = \frac{K_{FN2} \sigma_{FF2}}{S} = \frac{0.93 \times 380}{1.4} = 252.43 \text{ MPa}$$

IX. Calculate $\frac{Y_{Fa} Y_{Sa}}{[\sigma_F]}$.

$$\frac{Y_{Fa1} F_{Sa1}}{[\sigma_F]_1} = \frac{2.72 \times 1.57}{307.14} = 0.013,9, \quad \frac{Y_{Fa2} F_{Sa2}}{[\sigma_F]_2} = \frac{2.18 \times 1.79}{252.43} = 0.015,46$$

The value of the big gear is bigger.

(b) Calculation:

$$m_n \geq \sqrt[3]{\frac{2 \times 2.013,3 \times 4.051,8 \times 10^4 \times 0.75 \times \cos^2 14° \times 0.015,46}{1 \times 22^2 \times 1.645}} = 1.308 \text{ mm}$$

4) Determine the final dimension.

The normal module of tooth contact fatigue strength is greater than the normal module of tooth root bending fatigue strength according to the calculation results. So finally $m_n = 2$ mm, it have already meet the bending strength. We need to calculate the proper number of teeth by the pitch circle diameter $d_1 = 49.675,6$ mm in order to meet the contact fatigue strength, so

$$z_1 = \frac{49.675,6 \times \cos 14°}{m_n} = 21.422, \quad \text{so } z_1 = 24$$

$$z_2 = 4.448,2 \times 24 = 106.8, \quad \text{so } z_2 = 107$$

(a) Calculate the distance between the center of the two gears.

$$a = \frac{(z_1 + z_2)m_n}{2\cos\beta} = \frac{(24 + 107) \times 2}{2 \times \cos 14°} = 135.006 \text{ mm}$$

Round the center distance to 135 mm.

(b) Revise the helix angle by the rounded centre distance.

$$\beta = \arccos \frac{(z_1 + z_2)m_n}{2a} = \arccos \frac{(24 + 107) \times 2}{2 \times 135} = 13.982,3°$$

ε_α, k_β, Z_H don't need to be revised because its change is extremely small.

(c) Calculate pitch diameters of the two gears.

$$d_1 = \frac{z_1 m_n}{\cos\beta} = \frac{24 \times 2}{\cos 13.982°} = 49.46 \text{ mm}, \quad d_2 = \frac{z_2 m_n}{\cos\beta} = \frac{107 \times 2}{\cos 13.982°} = 220.53 \text{ mm}$$

Note: *here d_1 is a little smaller and it should be larger than* 49.675,6 mm.

(d) Calculate the width of the gear.

$$B = \varphi_d d_1 = 1 \times 49.46 = 49.46 \text{ mm}$$

Then $B_2 = 50$ mm, $B_1 = 55$ mm.

(2) Design and calculation of low-speed stage gear.

1) Basic parameters of the gears.

(a) Small and big gears of low-speed stage.

Small gear of low-speed stage:

Material: 40 Cr, tempering.

Rigidity: 280HBS.

Number of teeth: $z_1 = 30$.

Big gear of low-speed stage:

Material: 45 steel, tempering.

Rigidity: 240HBS.

Number of teeth: $z_2 = 3.294,9 \times 30 = 99$, choose $z_2 = 98$.

(b) Precision of the gear.

Choose grade 8 according to reference [1].

2) Calculation of the contact strength of gear.

Design by the tooth surface contact strength, according to the formula:

$$d_{1t} \geqslant \sqrt[3]{\frac{2K_t T_1}{\varphi_d \varepsilon_\alpha} \times \frac{u+1}{u} \times \left(\frac{Z_H Z_E}{\sigma_H}\right)^2}$$

(a) Identify the variables in the formula.

Chapter 20 Drawing Examples and Cases

I. Choose load factor:
$$K_t = 1.6$$

II. Choose zone factor:
$$Z_H = 2.433$$

III. According to the figure in reference [1],
$$\varepsilon_{a1} = 0.765, \quad \varepsilon_{a2} = 0.88 \Rightarrow \varepsilon_a = 0.765 + 0.88 = 1.645$$

IV. Torque of the small gear.
$$T_1 = 9.55 P_1/n_1 = 9.55 \times 3.872 \times 10^3 / 215.82 = 171.335 \text{ N}\cdot\text{m}$$

V. According to the figure in reference [1], the tooth width factor
$$\varphi_d = 1$$

VI. The elastic influence coefficient:
$$Z_E = 189.8 \text{ MPa}^{\frac{1}{2}}$$

VII. The contact strength limit of the small gear
$$\sigma_{Hlim1} = 600 \text{ MPa}$$

The contact strength limit of the big gear
$$\sigma_{Hlim1} = 550 \text{ MPa}$$

VIII. Calculate the number of stress cycles by the formula:
$$N_1 = 60 n_1 j L_h = 60 \times 215.82 \times 1 \times (2 \times 8 \times 300 \times 8) = 4.982,5 \times 10^8 \text{ h}$$
$$N_2 = 60 n_2 j L_h = 60 \times 65.5 \times 1 \times (2 \times 8 \times 300 \times 8) = 1.509 \times 10^8 \text{ h}$$

IX. Calculate the contact fatigue allowable stress.

Safety factor $S=1$. By the formula:
$$[\sigma_H]_1 = \frac{K_{HN1} \sigma_{Hlim1}}{S} = \frac{0.94 \times 600}{1} = 564 \text{ MPa}$$

$$[\sigma_H]_2 = \frac{K_{HN2} \sigma_{Hlim2}}{S} = 0.98 \times 550/1 = 517 \text{ MPa} \Rightarrow$$

$$[\sigma_H] = ([\sigma_H]_1 + [\sigma_H]_2)/2 = (564 + 517)/2 = 540.5 \text{ MPa}$$

(b) Calculate.

I. Calculate the pitch diameter of the small gear.
$$d_{1t} \geq \sqrt[3]{\frac{2 K_t T_1}{\varphi_d \varepsilon_a} \times \frac{u+1}{u} \times \left(\frac{Z_H Z_E}{[\sigma_H]}\right)^2} =$$

$$\sqrt[3]{\frac{2 \times 1.6 \times 171.335 \times 10^3}{1 \times 1.645} \times \frac{4.448,2}{3.448,2} \times \left(\frac{2.433 \times 189.8}{540.5}\right)^2} = 67.957 \text{ mm}$$

II. Calculate the circular velocity.
$$v = \frac{\pi d_{1t} n_2}{60 \times 1,000} = \frac{\pi \times 67.957 \times 215.82}{60 \times 1,000} = 0.767,5 \text{ m/s}$$

III. Calculation of tooth width b and module m_{nt}.
$$b = \varphi_d \times d_{1t} = 67.957 \text{ mm}$$

$$m_{nt} = \frac{d_{1t}\cos\beta}{z_1} = \frac{67.957 \times \cos14°}{30} = 2.198 \text{ mm}$$

$$h = 2.25 m_{nt} = 2.25 \times 2.198 = 4.945,5 \text{ mm}$$

$$\frac{b}{h} = b/h = \frac{67.957}{4.945,5} = 13.741$$

IV. Calculation of vertical overlap contact ratio.

$$\varepsilon_\beta = 0.318\varphi_d z_1 \tan\beta = 0.318 \times 30 \times \tan14° = 2.378,6$$

V. Calculation of the load factor K.

$$K_{H\beta} = 1.15 + 0.18\varphi_d^2 + 0.31 \times 10^{-3} b = 1.15 + 0.18 + 0.31 \times 10^{-3} \times 67.957 = 1.351$$

According to the table in reference [1], $K_A = 1$.

The same to the design of the high-speed stage gear:

$$K_v = 1.05, \quad K_{F\beta} = 1.30, \quad K_{H\alpha} = K_{F\alpha} = 1.4$$

So the load factor

$$K = K_A K_v K_{H\alpha} K_{H\beta} = 1 \times 1.05 \times 1.30 \times 1.351 = 1.844$$

VI. Modify the pitch circle diameter by the actual load factor:

$$d_1 = d_{1t} \sqrt[3]{K/K_t} = 67.957 \times \sqrt[3]{\frac{1.844}{1.6}} = 71.27 \text{ mm}$$

VII. Calculate the modulus:

$$m_n = \frac{d_1 \cos\beta}{z_1} = \frac{71.27 \times \cos14°}{30} = 2.305 \text{ mm}$$

3) Calculation of the bending strength of gear.

Calculate by formula:

$$m_n \geqslant \sqrt[3]{\frac{2KT_1 Y_\beta \cos^2\beta}{\varphi_d Z_1^2 \varepsilon_a} \left(\frac{Y_{Fa} Y_{Sa}}{[\sigma_F]}\right)}$$

(a) Determine the parameters in the formula.

I. Calculate load factor.

$$K = K_A K_v K_{F\alpha} K_{F\beta} = 1 \times 1.05 \times 1.4 \times 1.30 = 1.947,4$$

II. According to the overlap contact ratio, $\varepsilon_\beta = 2.378,6$, we get $Y_\beta = 0.75$,

III. Calculate equivalent number of teeth.

$$z_{v1} = z_1/\cos^3\beta = 30/\cos^3 14° = 32.84$$

$$z_{v2} = z_2/\cos^3\beta = 99/\cos^3 14° = 108.37$$

IV. Choose form factor from the table in reference [1], $Y_{Fa1} = 2.52$, $Y_{Fa2} = 2.18$.

V. Choose stress correction factor from the table in reference [1], $Y_{Sa1} = 1.625$, $Y_{Sa2} = 1.79$.

VI. Choose the bending fatigue limit from figure.

$$\sigma_{FE1} = 500 \text{ MPa}, \quad \sigma_{FE2} = 380 \text{ MPa}$$

VII. Choose bending fatigue life factor from figure.

$$K_{FN1} = 0.90, \quad K_{FN2} = 0.93$$

Chapter 20 Drawing Examples and Cases

Ⅷ. Calculate the allowable bending stress fatigue, choose bending fatigue safety factor $S=1.4$,

$$[\sigma_F]_1 = \frac{K_{FN1}\sigma_{FE1}}{S} = \frac{0.90 \times 500}{1.4} = 321.43 \text{ MPa}$$

$$[\sigma_F]_2 = \frac{K_{FN2}\sigma_{FF2}}{S} = \frac{0.93 \times 380}{1.4} = 252.43 \text{ MPa}$$

Ⅸ. Calculate $\dfrac{Y_{Fa}Y_{Sa}}{[\sigma_F]}$.

$$\frac{Y_{Fa1}Y_{Sa1}}{[\sigma_F]_1} = \frac{2.52 \times 1.625}{321.43} = 0.012,74, \quad \frac{Y_{Fa2}Y_{Sa2}}{[\sigma_F]_2} = \frac{2.18 \times 1.79}{252.43} = 0.015,46$$

The value of the big gear is bigger.

(b) Calculation.

$$m_n \geq \sqrt[3]{\frac{2 \times 1.947,4 \times 1.713,35 \times 10^5 \times 0.75 \times \cos^2 14° \times 0.015,46}{1 \times 30^2 \times 1.645}} = 1.700,85 \text{ mm}$$

4) Determine the final dimension.

The normal module of tooth contact fatigue strength is greater than the normal module of tooth root bending fatigue strength according to calculation results. So choose $m_n = 2.5$ mm, it has already met the bending strength. We need to calculate the proper number of teeth by the pitch circle diameter $d_1 = 71.27$ mm in order to meet the contact fatigue strength, so

$$z_1 = \frac{71.27 \times \cos 14°}{m_n} = 27.66, \quad \text{choose } z_1 = 27$$

$$z_2 = 3.294,9 \times 27 = 88.9, \quad \text{choose } z_2 = 89$$

(a) Calculate the distance between the centre of the two gears.

$$a = \frac{(z_1 + z_2)m_n}{2\cos\beta} = \frac{(27+89) \times 2.5}{2 \times \cos 14°} = 149.439 \text{ mm}$$

Round the center distance to 150 mm.

(b) Revise the helix angle by the rounded centre distance.

$$\beta = \arccos\frac{(z_1+z_2)m_n}{2a} = \arccos\frac{(27+89) \times 2.5}{2 \times 150} = 14.835°$$

ε_α, k_β, Z_H don't need to be revised for its small change.

(c) Calculate the pitch diameters of the two gears.

$$d_1 = \frac{z_1 m_n}{\cos\beta} = \frac{27 \times 2.5}{\cos 14.835°} = 69.83 \text{ mm}$$

$$d_2 = \frac{z_2 m_n}{\cos\beta} = \frac{89 \times 2.5}{\cos 14.835°} = 230.17 \text{ mm}$$

Note: here d_1 should be larger than 71.27 mm.

(d) Calculate the width of the gear.

$$b = \varphi_d d_1 = 1 \times 69.83 = 69.83 \text{ mm}$$

After being rounded, $B_1=70$ mm, $B_2=75$ mm.

5. Design and calculation of shaft

(1) Design of high-speed shaft.

1) Structural design of high-speed shaft.

(a) Estimate the minimum diameter.

According to the table in reference [1], $A_0=112$.

We can calculate the minimum diameter of shaft:

$$d_{min}=A_0\sqrt[3]{\frac{P_1}{n_1}}=112\sqrt[3]{\frac{4.073}{960}}=18.13 \text{ mm}$$

The segment of the shaft with minimum diameter is connecting with the coupling.

(b) Determine the segment's sizes of the shaft.

According to the requirements of axial location, determine the diameter and length of each segment (see Fig. 20-7).

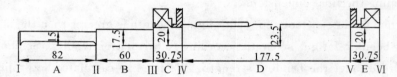

Fig. 20-7 Dimensions of high-speed shaft

Ⅰ. Determine the shaft diameter.

Segment A: $d_1=30$ mm. The diameter of the half coupling should be considered.

Segment B: $d_2=35$ mm, according to the standard of seal.

Segment C: $d_3=40$ mm. To fit the bearing (tapered roller bearings 30208), $d_3=$ bearing inner diameter.

Segment D: $d_4=47$ mm. The locating shoulder height is 3.5 mm, and here a gear shaft is designed.

Segment E: $d_5=40$ mm. To fit the bearing (tapered roller bearings 30208), $d_3=d_5$.

Ⅱ. Determine the length of each segment.

Segment A: $L_1=82$ mm. The length of the half coupling is 82 mm (L_1 *should be a little smaller than* 82 mm, *and* 80 mm *is suggested*).

Segment B: $L_2=60$ mm. The width of bearing cover and the distance between bearing cover and the coupling are considered.

Segment C: $L_3=30.75$ mm. The bearing (tapered roller bearings 30208) and oil baffle are installed here.

Segment D: $L_4=177.5$ mm. The distance between gear and housing $a=16$ mm, the distance between two gears $c=20$ mm. Then according to the widths of gears on the middle shaft, L_4 is finally determined as 177.5 mm.

Segment E: $L_5 = 30.75$ mm. The bearing(tapered roller bearings 30208)and oil baffle are installed here.

2)Strength check of the high-speed shaft.

(a)Stress analysis of shaft.

According to tapered roller bearings 30208, we get the distance between the center of the gear and the bearing as shown in Fig. 20 - 8.

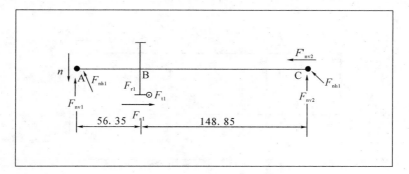

Fig. 20 - 8 Force analysis of high speed shaft

Ⅰ. Known: the pitch diameter of the high-speed gear.

$$d_1 = 47.46 \text{ mm}$$

$$F_{t1} = \frac{2T_1}{d_1} = \frac{2 \times 40.518}{47.46 \times 10^3} = 1,707.46 \text{ N}$$

$$F_{r1} = \frac{\tan\alpha_n}{\cos\beta} \cdot F_{t1} = \frac{\tan 20°}{\cos 14.25°} \times 1,707.46 = 641.19 \text{ N}$$

$$F_{a1} = F_{t1}\tan\beta = 1,707.46 \times \tan 14.25° = 433.64 \text{ N}$$

Ⅱ. Calculate the supporting force.

On the horizontal plane(see Fig. 20 - 9):

$$F_{nh1} = \frac{148.85}{56.35 + 148.85}F_{t1} = 1,707.46 \times \frac{148.85}{205.2} = 1,238.57 \text{ N}$$

$$F_{nh1} = \frac{56.35}{56.35 + 148.85}F_{t1} = 1,707.46 \times \frac{56.35}{205.2} = 468.88 \text{ N}$$

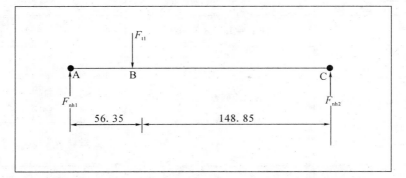

Fig. 20 - 9 Force analysis of high speed shaft on horizontal plane

On the vertical plane (see Fig. 20-10):

$$F_{nv1} = \frac{F_{r1} \times 148.85 + \frac{F_{a1}d_1}{2}}{56.35 + 148.85} = \frac{641.19 \times 148.85 + \frac{433.64 \times 47.46}{2}}{205.2} = 515.26 \text{ N}$$

$$F_{nv2} = F_{r1} - F_{nv1} = 641.19 - 515.26 = 125.93 \text{ N}$$

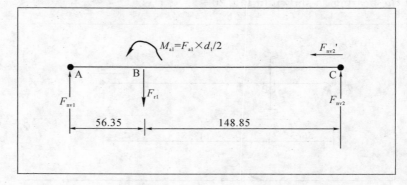

Fig. 20-10 Force analysis of high speed shaft on vertical plane

(b) Check calculation according to the condition of bending and torsion.

$$M_h = F_{nh1} \times 56.35/10^3 = 69.79 \text{ N}$$
$$M_{v1} = F_{nv1} \times 56.35/10^3 = 29.03 \text{ N}$$
$$M_{v2} = F_{nv2} \times 148.85/10^3 = 18.74 \text{ N}$$
$$M_1 = \sqrt{M_h^2 + M_{v1}^2} = \sqrt{69.79^2 + 29.03^2} = 75.587 \text{ N}$$
$$M_2 = \sqrt{M_h^2 + M_{v2}^2} = \sqrt{69.79^2 + 18.74^2} = 72.26 \text{ N}$$
$$T_1 = 40.518 \text{ N} \cdot \text{m}$$

The results are listed in the following Table 20-3. *The calculated values are not completely correct.*

Table 20-3 Results of force, moment and torque

Load	Horizon(H)	Vertical(V)
Supporting force F	$F_{nh1} = 1,238.57$ N	$F_{nv1} = 515.26$ N
	$F_{nh2} = 468.88$ N	$F_{nv2} = 125.93$ N
Bending moment M	$M_h = 69.79$ N·m	$M_{v1} = 29.03$ N·m $M_{v2} = 18.74$ N·m
The total bending moment	$M = \sqrt{69.79^2 + 29.03^2} = 75.587$ N·m	
Torque T	$T_1 = 40.518$ N·m	

The bending moment, torque diagram are shown in Fig. 20-11.

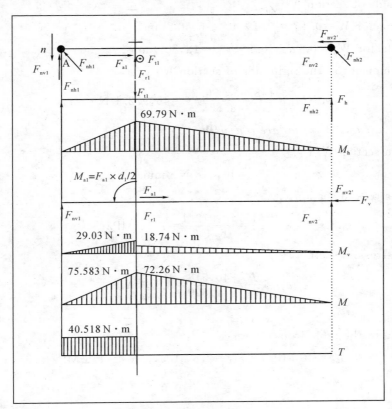

Fig. 20-11 Bending moment and torque diagram

Check calculation according to the condition of bending and torsion.

According to the above formula and the data table, we can choose $a=0.6$. Calculation stress of the shaft:

$$\sigma_{ca}=\frac{\sqrt{M^2+(\alpha T_1)^2}}{W}=\frac{\sqrt{75,587^2+(0.6\times 40,518)^2}}{0.1\times 42.46^3}=9.88 \text{ MPa}$$

The material of the shaft that we have selected is 45 steel, which is hardened and tempered. According to the required table, get $[\sigma_{-1}]=60$ MPa, $\sigma_{ca}<[\sigma_{-1}]$. So it meets the requirements.

(c) Accurate check by fatigue.

I. Determine the dangerous section.

Only by the torque, sections I, II, III don't need to be checked. Because of the impact of stress concentration on the shaft, the stress concentration caused by sections V and IV is the most serious. Section VI doesn't need to be checked for its big diameter and without torque. Although the stress in the middle surface of the gear is the maximum, it doesn't have to be checked for its big diameter. So we only need to check both sides of section Ⅳ.

Ⅱ. Left side of section Ⅳ.

Bending factor: $W = 0.1d^3 = 0.1 \times 40^3 = 6,400 \text{ mm}^3$.

Torsion factor: $W_T = 0.2d^3 = 0.2 \times 40^3 = 12,800 \text{ mm}^3$.

Bending moment in the right side of section Ⅳ:

$$M = M_1 \times \frac{148.85 - 56.35}{148.85} = 46,972.1 \text{ N} \cdot \text{mm}$$

Here the calculated values are confused.

Torque in section Ⅳ T:

$$T = 40,518 \text{ N} \cdot \text{mm}.$$

Bending stress in section:

$$\sigma_b = \frac{M}{W} = \frac{46,972.1}{6,400} = 7.34 \text{ MPa}$$

Torsion stress in section:

$$\sigma_T = \frac{T}{W_T} = \frac{40,518}{12,800} = 3.165 \text{ MPa}$$

Material is 45 steel.

According to the table from reference [1]:

$$\sigma_b = 640 \text{ MPa}, \quad \sigma_{-1} = 275 \text{ MPa}, \quad \tau_{-1} = 155 \text{ MPa}$$

Because

$$\frac{r}{d} = \frac{2.0}{40} = 0.05$$

$$\frac{D}{d} = \frac{47}{40} = 1.175$$

According to the table in reference [1]:

$$\alpha_\sigma = 2.09, \quad \alpha_\tau = 1.66$$

The factor:

So

$$q_\sigma = 0.82, \quad q_\tau = 0.85$$

$$k_\sigma = 1 + q_\sigma(\alpha_\sigma - 1) = 1 + 0.82 \times (2.09 - 1) = 1.89$$

$$k_\tau = 1 + q_\tau(\alpha_\tau - 1) = 1 + 0.85 \times (1.66 - 1) = 1.56$$

$$\varepsilon_\sigma = 0.82, \quad \varepsilon_\tau = 0.85, \quad \beta_\sigma = \beta_\tau = 0.92$$

Comprehensive coefficient:

$$K_\sigma = \frac{k_\sigma}{\varepsilon_\sigma} + \frac{1}{\beta_\sigma} - 1 = \frac{1.89}{0.82} + \frac{1}{0.92} - 1 = 2.6$$

$$K_\tau = \frac{k_\tau}{\varepsilon_\tau} + \frac{1}{\beta_\tau} - 1 = \frac{1.561}{0.86} + \frac{1}{0.92} - 1 = 1.9$$

Coefficient of carbon steel:

$$\varphi_\sigma = 0.1 - 0.2, \quad \varphi_\sigma = 0.1$$

$$\varphi_\tau = 0.05 - 0.1, \quad \varphi_\tau = 0.05$$

Chapter 20 Drawing Examples and Cases

Safety factor S_{ca}:

$$S_\sigma = \frac{\sigma_{-1}}{K_\sigma \sigma_a + \varphi_\sigma \sigma_m} = \frac{275}{1.89 \times 7.34 + 0.1 \times 0} = 19.82$$

$$S_\tau = \frac{\tau_{-1}}{K_\tau \tau_a + \varphi_\tau \tau_m} = \frac{155}{1.561 \times \frac{3.165}{2} + 0.05 \times \frac{3.165}{2}} = 60.8$$

$$S_{ca} = \frac{S_\sigma S_\tau}{\sqrt{S_\sigma^2 + S_\tau^2}} = \frac{19.82 \times 60.8}{\sqrt{19.82^2 + 60.8^2}} = 18.84 \geqslant S = 1.5$$

So it meets the requirements.

Ⅲ. Right side of section Ⅳ.

Bending factor: $W = 0.1 d^3 = 0.1 \times 47^3 = 10,382.3 \text{ mm}^3$.

Torsion factor: $W_T = 0.2 d^3 = 0.2 \times 47^3 = 20,764.6 \text{ mm}^3$.

Bending moment in the right side of section:

$$M = M_1 \times \frac{148.85 - 56.35}{148.85} = 46,972.1 \text{ N} \cdot \text{mm}$$

Torque in section Ⅴ T:

$$T = 40,518 \text{ N} \cdot \text{mm}$$

Bending stress in section:

$$\sigma_b = \frac{M}{W} = \frac{46,972.1}{10,382.3} = 4.52 \text{ MPa}$$

Torsion stress in section:

$$\sigma_T = \frac{T}{W_T} = \frac{46,972.1}{20,764.6} = 2.26 \text{ MPa}$$

Material is 45 steel.

According to the table from reference [1]:

$\sigma_B = 640 \text{ MPa}$, $\sigma_{-1} = 275 \text{ MPa}$, $\tau_{-1} = 155 \text{ MPa}$, $\beta_\sigma = \beta_\tau = 0.92$

Comprehensive coefficient:

$$K_\sigma = \frac{k_\sigma}{\varepsilon_\sigma} + \frac{1}{\beta_\sigma} - 1 = 2.3 + \frac{1}{0.92} - 1 = 2.39$$

$$K_\tau = \frac{k_\tau}{\varepsilon_\tau} + \frac{1}{\beta_\tau} - 1 = 1.84 + \frac{1}{0.92} - 1 = 1.93$$

Coefficient of carbon steel:

$$\varphi_\sigma = 0.1 - 0.2, \quad \varphi_\sigma = 0.1$$

$$\varphi_\tau = 0.05 - 0.1, \quad \varphi_\tau = 0.05$$

Safety factor S_{ca}:

$$S_\sigma = \frac{\sigma_{-1}}{K_\sigma \sigma_a + \varphi_\sigma \sigma_m} = \frac{275}{2.39 \times 2.17 + 0.1 \times 0} = 53.02 \text{ MPa}$$

$$S_\tau = \frac{\tau_{-1}}{K_\tau \sigma_a + \varphi_\tau \tau_m} = \frac{155}{1.93 \times \frac{2.26}{2} + 0.05 \times \frac{2.26}{2}} = 69.3 \text{ MPa}$$

$$S_{ca}=\frac{S_\sigma S_\tau}{\sqrt{S_\sigma^2+S_\tau^2}}=\frac{53.02\times 69.3}{\sqrt{53.02^2+69.3^2}}=42.1\geqslant S=1.5$$

So it meets the requirements.

Note: *some of the calculations are not correct.*

(2) Design of the middle shaft.

(3) Design of the low-speed shaft.

Using the similar method, the middle shaft and the low shaft can be designed and calculated.

6. Selection and calculation of bearing life

(1) Check calculation of high-speed shaft bearings.

Choose single-row tapered roller bearing 30208 (see Fig. 20-12).

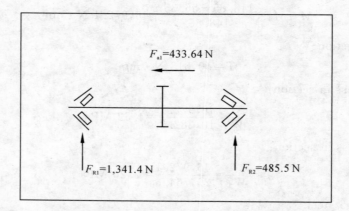

Fig. 20-12 Force analysis of high-speed shaft bearings

Its parameter:

$d\times D\times T=30$ mm$\times 72$ mm$\times 20.75$ mm, $F_{a1}=433.64$ N, $C_r=59.8$ kN, $X=0.4$, $Y=1.6$.

From Table 20-3, we get $F_{nh1}=1,238.5$ N, $F_{nv1}=515.26$ N, $F_{nh2}=468.88$ N, $F_{nv2}=125.93$ N.

So $F_{R1}=\sqrt{F_{nh1}^2+F_{nv1}^2}=1,341.4$ N, $F_{R_2}=\sqrt{F_{nh2}^2+F_{nv2}^2}=485.5$ N

So
$$F_{d1}=\frac{F_{R1}}{2Y}=\frac{1,341.4}{2\times 1.6}=419.2\text{ N}$$

$$F_{d2}=\frac{F_{R2}}{2Y}=\frac{485.5}{2\times 1.6}=151.72\text{ N.}$$

Left force: $F_{d2}+F_{a1}=585.36$ N; right force: $F_{d1}=419.2$ N.

Left force > right force.

So $F_{A1}=F_{a1}+F_{d2}=585.36$ N, $F_{A2}=F_{d2}=151.72$ N.

Equivalent dynamic load of bearing P_1 and P_2:

$$P_1 = f_p(XF_{R1} + YF_{A1}) = 1 \times (0.4 \times 1314.4 + 1.6 \times 585.36) = 1,473.1 \text{ N}$$
$$P_2 = f_p(XF_{R2} + YF_{A2}) = 1 \times (0.4 \times 485.5 + 1.6 \times 151.72) = 436.95 \text{ N}$$

Because $\quad P_1 > P_2$

$$C = P_1 \sqrt[3]{\frac{60 n_1 L_h'}{10^6}} = 1,473.1 \times \sqrt[3]{\frac{60 \times 960 \times 2 \times 8 \times 300 \times 8}{10^6}} = 19.193 \text{ kN}$$

Because $\quad C < C_r = 59.8 \text{ kN}$

So 30208 meets the requirements.

(2) Check calculation of middle shaft bearings.

Choose single-row tapered roller bearing 30207 (see Fig. 20-13).

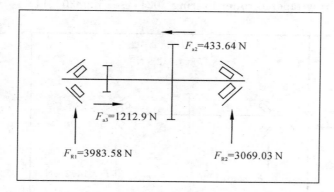

Fig. 20-13 Force analysis of middle shaft bearings

Its parameter:

$d \times D \times T = 35 \text{ mm} \times 72 \text{ mm} \times 18.25 \text{ mm}$, $F_{a2} = 433.64 \text{ N}$, $F_{a3} = 1,212.9 \text{ N}$, $C_r = 51.5 \text{ kN}$, $X = 0.4$, $Y = 1.6$.

From Table 20-4, we get $F_{nh1} = 3,567.1 \text{ N}$, $F_{nv1} = 1,773.34 \text{ N}$, $F_{nh2} = 2,719.67 \text{ N}$, $F_{nv2} = 1,422.1 \text{ N}$.

So
$$F_{R1} = \sqrt{F_{nh1}^2 + F_{nv1}^2} = 3,983.58 \text{ N}$$
$$F_{R2} = \sqrt{F_{nh2}^2 + F_{nv2}^2} = 3,069.03 \text{ N}$$

So
$$F_{d1} = \frac{F_{R1}}{2Y} = \frac{3,983.58}{2 \times 1.6} = 1,244.87 \text{ N}$$
$$F_{d2} = \frac{F_{R2}}{2Y} = \frac{3,069.03}{2 \times 1.6} = 959.07 \text{ N}$$

Left force: $F_{d2} + F_{a2} = 433.64 + 959.07 = 1,392.71 \text{ N}$.

Right force: $F_{d1} + F_{a3} = 1,244.87 + 1,212.9 = 2,457.77 \text{ N}$.

Left force < Right force.

So $\quad F_{A1} = F_{d1} = 1,244.87 \text{ N}$

$$F_{A2}=F_{d1}+F_{a3}-F_{a2}=1,244.87+1,212.9-433.64=2,024.13 \text{ N}$$

Equivalent dynamic load of bearing, P_1 and P_2:

$$P_1=f_p(XF_{R1}+YF_{A1})=1\times(0.4\times3,983.58+1.6\times1,244.87)=3,585.2 \text{ N}$$
$$P_2=f_p(XF_{R2}+YF_{A2})=1\times(0.4\times3,069.03+1.6\times2,024.13)=4,466.22 \text{ N}$$
$$P_1<P_2$$
$$C=P_2\sqrt[3]{\frac{60n_1L'_h}{10^6}}=4,466.22\times\sqrt[3]{\frac{60\times215.82\times2\times8\times300\times8}{10^6}}=35.383 \text{ kN}$$

Because
$$C<C_r=51.5 \text{ kN}$$

30207 meets the requirements.

(3) Check calculation of low-speed shaft bearings.

Choose single-row tapered roller bearing 30211(see Fig. 20 - 14).

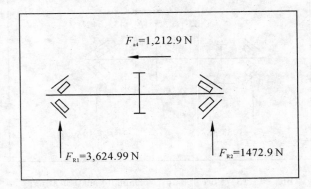

Fig. 20 - 14　Force analysis of low speed shaft bearings

Its parameter:

$d\times D\times T=55$ mm$\times100$ mm$\times22.75$ mm, $F_{a4}=433.64$ N, $C_r=86.5$ kN, $X=0.4$, $Y=1.5$.

The calculation procedure is the same with that of high speed shaft and middle shaft. It meets the requirements.

7. Selection and checking calculation of key

(1) High-speed shaft.

1) Key dimensions.
$$b\times h\times L=8 \text{ mm}\times7 \text{ mm}\times70 \text{ mm}$$

2) Checking calculation of key.

$k=0.5h=0.5\times7=3.5$ mm, $l=L-b/2=70-4=66$ mm, $d=30$ mm

$\sigma_p=2T/kld=2\times40.518\times10^3/(3.5\times66\times30)=11.7$ MPa$\leqslant[\sigma_p]=110$ MPa

So it meets the requirements.

Chapter 20　Drawing Examples and Cases

(2) Middle shaft

1) Key dimensions.
$$b \times h \times L = 12 \text{ mm} \times 8 \text{ mm} \times 45 \text{ mm}$$

2) Checking calculation of key.
$$k = 0.5h = 0.5 \times 8 = 4 \text{ mm}, \quad l = L - b = 45 - 12 = 33 \text{ mm}, \quad d = 44 \text{ mm}$$
$$\sigma_p = 2T/kld = 2 \times 171.355 \times 10^3 / (4 \times 33 \times 44) = 59 \text{ MPa} \leqslant [\sigma_p] = 110 \text{ MPa}$$

So it meets the requirements.

(3) Low-speed shaft.

1) Key dimensions.
$$b \times h \times L = 14 \text{ mm} \times 9 \text{ mm} \times 100 \text{ mm}$$

2) Checking calculation of key.
$$k = 0.5h = 0.5 \times 9 = 4.5 \text{ mm}, \quad l = L - b/2 = 100 - 7 = 93 \text{ mm}, \quad d = 45 \text{ mm}$$
$$\sigma_p = 2T/kld = 2 \times 536.695 \times 10^3 / (4.5 \times 96 \times 45) = 36.6 \text{ MPa} \leqslant [\sigma_p] = 110 \text{ MPa}$$

So it meets the requirements.

8. Selection of coupling

(1) Coupling of the input shaft.

According to the table in reference [1], thinking of the small relatively range of the torque, $K_A = 1.3$, $T_{ca} = K_A T_3 = 1.3 \times 40.518 = 52.67$ N·m.

Choose ML3 plum-type flexible coupling for the input shaft.

Its parameter:

nominal torque $T = 90$ N·m,　$d_1 = 30$ mm,　$L_1 = 82$ mm.

So it meets the requirements.

(2) Coupling of the output shaft.

According to the table in reference [1], thinking of the small relatively range of the torque, $K_A = 1.3$, $T_{ca} = K_A T_3 = 1.3 \times 536.695 = 697.7$ N·m.

Choose TL8 elastic sleeve pin coupling for the output shaft.

Its parameter:

nominal torque: $T = 710$ N·m,　$L_1 = 112$ mm,　$d_1 = 45$ mm.

So it meets the requirements.

9. Lubrication

For the two cylindrical gear reducer, because the gear is lightweight and low speed, its speed $v = 1.83$ m/s < 2 m/s, we should use grease to lubricate the bearings.

Oil was used oil to lubricate the gear, according to the table from reference [1], and industrial gear oil N220 was selected.

Look at Fig. 20-15 Carefully, it can be found that the layout of the gears is different from that of Fig. 20-6. Generally high-speed level gear should be located far away from the input of the torque. So this reducer is not designed rationally.

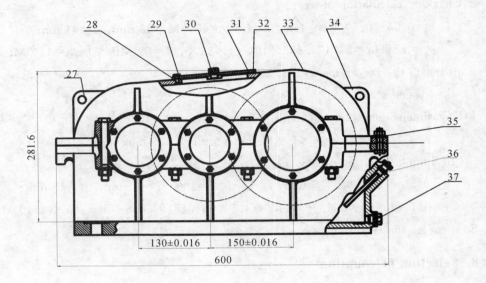

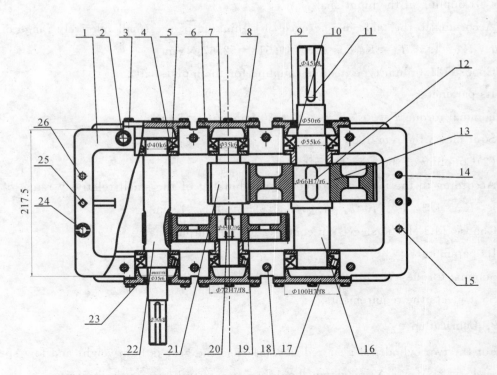

Fig. 20-15 Two cylindrical gear reducer

Chapter 20 Drawing Examples and Cases

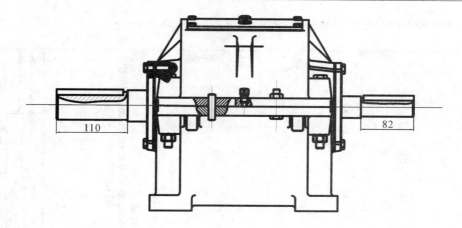

Technical characteristics

Input power/kW	High-speed shaft speed/(r·min^{-1})	Eefficiency η	Drive ratio i
5.5	960	0.811,6	14.656,5

Technical requirememts

1. Before the assembly, the unmachined surface of the housing and other casting parts should be cleaned, burrs should be removed, and antirust liquid should beused.

2. Before the assembly, components should be cleaned by kerosene, bearings by gasoling cleaning, after drying the fit surface should be oiled.

3. Oil leakage should be avoided on the spilt surface, contacting surface and sealing surface, sealant or water glass can be used on the splip surface and other filler should not be used.

4. After assembly, contace pits should be checked by coloring method, along gear height not more 30%, along gear length not more than 50%, the backlash:the first grade jnmin=0.160 mm.

5. When adjust and fix bearings, there should be an axial clearance of 0.2 - 0.5 mm.

6. The gear reducer is equipped with 220 industrial gear oil, oil reaches the specified depth.

7. The inside wall should be painted with anti oil paint, reducer surface painted with grey paint.

8. Experiments should be done according to the experiment procedure.

Continued Fig. 20 - 15 Two cylindrical gear reducer

The low-speed shaft and the larger gear on it are shown in Fig. 20 – 16 and Fig. 20 – 17.

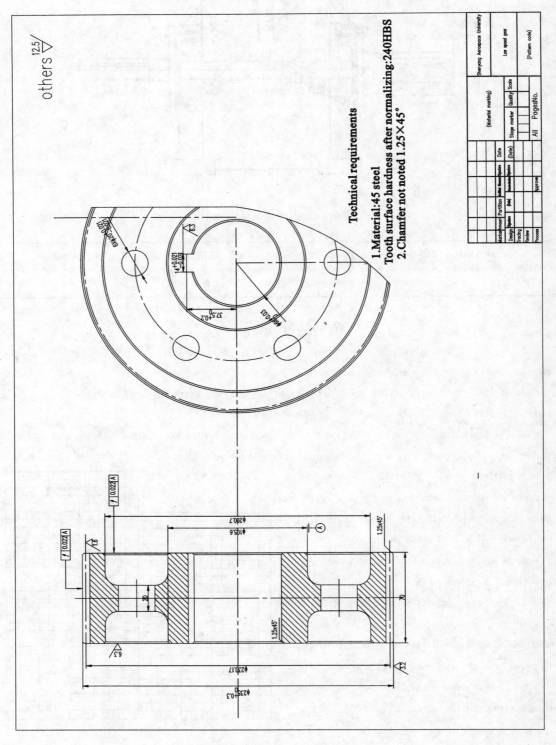

Fig. 20 – 16 Gear

Chapter 20　Drawing Examples and Cases

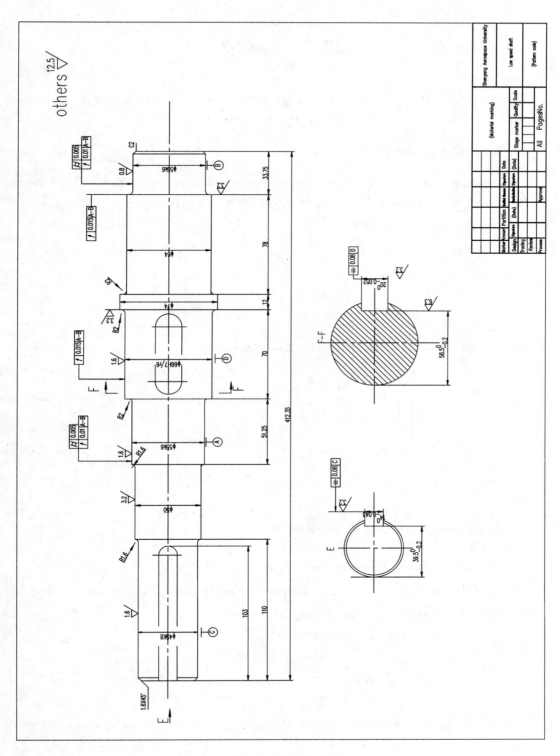

Fig. 20 - 17　Shaft

20.3 Case 2

This is completed by another student, the design and calculations are not completely correct, either.

20.3.1 Design data

As shown in Fig. 20-18, The reducer is required to work continuously, without reverse under stable load and clean environment, for 8 years (16 h per day). The power of the working part is 1.95 kW and the speed of the working part is 60 r/min.

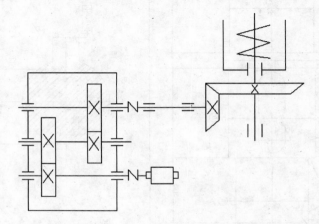

Fig. 20-18 Transmission of a conveyor

20.3.2 Design procedure and results

1. Selection and calculation of motor

Choose three phase asynchronous motor according to the working conditions and job requirements. Open structure, power 380 V, Y series.

Choose the power of working part:

$$P_W = 1.95 \text{ kW}$$

The total transmission efficiency:

$$\eta = \eta_{gear}^2 \cdot \eta_{bearing}^5 \cdot \eta_{flexible\text{-}coupling}^2 \cdot \eta_{bevelgear}$$

There are 5 pairs of bearings, 2 pairs of couplings, 1 opening bevel gear, and 2 pairs of closed gear.

Here the efficiency of the working part is canceled, but it is suggested to be considered.

According to Table 10-1, choose:

Efficiency of closed gear meshing:
$$\eta_{gear} = 0.97 \text{ (accuracy grade: 8)}$$

Efficiency of bearing:
$$\eta_{bearing} = 0.99 \text{ (angular contact ball bearing)}$$

Efficiency of coupling:
$$\eta_{flexible-bearing} = 0.99$$

Efficiency of the opening bevel gear:
$$\eta_{bevel\ gear} = 0.95$$

The total transmission efficiency:
$$\eta = 0.99^2 \times 0.99^5 \times 0.97^2 \times 0.95 = 0.83$$

The power required:
$$P_r = P_w/\eta = 1.95/0.83 = 2.35 \text{ kW}$$

According to Table 19-1, choose three phase asynchronous motor model: Y100L2-4, whose rated power: $P_0 = 3$ kW, full load speed $n = 1,420$ r/min and synchronous speed = 1,500 r/min.

2. Distribution of drive ratio

Total speed ratio $\quad i_t = n_{motor}/n_W = 1,420/60 = 24$.

The transmission ratio of the bevel gear $\quad i_{bevel\ gear} = 2.1$.

The transmission ratio of the reducer $\quad i_{reducer} = 11.43$.

Take the high-level gear transmission ratio:
$$i_{12} = \sqrt{1.3 \times i_{reeducer}} = 3.9$$

So the low-level gear transmission ratio:
$$i_{34} = i_{reducer}/i_{12} = 11.43/3.9 \approx 3$$

Generally the calculated values are not suggested to be rounded off.

3. Selection and calculation of kinematic and dynamic parameters

(1) Parameters of the motor shaft (0 shaft).
$$P_0 = P_r = 2.35 \text{ kW}$$
$$n_0 = 1,420 \text{ r/min}$$
$$T_0 = \frac{9.55 P_0}{n_0} = 9.55 \times 2.35 \times 10^3/1,420 = 15.80 \text{ N} \cdot \text{m}$$

(2) Parameters of the high-speed shaft (Ⅰ shaft).
$$P_1 = P_0 \eta_{coupling} = 2.35 \times 0.99 = 2.33 \text{ kW}$$
$$n_1 = n_0 = 1,420 \text{ r/min}$$
$$T_1 = \frac{9.55 P_1}{n_1} = 9.55 \times 2.33 \times 10^3/1,420 = 15.67 \text{ N} \cdot \text{m}$$

(3) Parameters of the middle shaft (Ⅱ shaft).

$$P_2 = P_1 \eta_{gear} \eta_{bearing} = 2.33 \times 0.97 \times 0.99 = 2.24 \text{ kW}$$

$$n_2 = \frac{n_1}{i_{12}} = \frac{1,420}{3.9} = 364.10 \text{ r/min}$$

$$T_2 = \frac{9.55 P_2}{n_2} = 9.55 \times 2.24 \times 10^3 / 364.10 = 58.75 \text{ N} \cdot \text{m}$$

(4) Parameters of the low-speed shaft (Ⅲ shaft).

$$P_3 = P_2 \eta_{gear} \eta_{bearing} = 2.24 \times 0.97 \times 0.99 = 2.15 \text{ kW}$$

$$n_3 = \frac{n_2}{i_{34}} = \frac{364.10}{3} = 121.37 \text{ r/min}$$

$$T_3 = \frac{9.55 P_3}{n_3} = 9.55 \times 2.15 \times 10^3 / 121.37 = 169.17 \text{ N} \cdot \text{m}$$

(5) Parameters of the fourth shaft (Ⅳ shaft).

$$P_4 = P_3 \eta_{couplling} \eta_{bearing} = 2.15 \times 0.99 \times 0.99 = 2.11 \text{ kW}$$

$$n_4 = n_3 = 121.37 \text{ r/min}$$

$$T_4 = \frac{9.55 P_4}{n_4} = 9.55 \times 2.11 \times 10^3 / 121.37 = 166.03 \text{ N} \cdot \text{m}$$

(6) Parameters of the fifth shaft (Ⅴ shaft).

$$P_5 = P_4 \eta_{bevel\ gear} \eta_{bearing} = 2.11 \times 0.95 \times 0.99 = 1.98 \text{ kW}$$

$$n_5 = \frac{n_4}{i_{bevel\ gear}} = 121.37 / 2.1 = 57.80 \text{ r/min}$$

$$T_5 = \frac{9.55 P_5}{n_5} = 9.55 \times 1.98 \times 10^3 / 57.80 = 327.15 \text{ N} \cdot \text{m}$$

The results summarized above are listed in Table 20 – 4.

Table 20 – 4 **Kinematic and dynamic parameters of each shaft**

Shaft No.	Power /kW	Rotating speed /(r · min^{-1})	Torque /(N · m)
0	2.35	1420	15.80
Ⅰ	2.33	1420	15.67
Ⅱ	2.24	364.10	58.75
Ⅲ	2.15	121.37	169.17
Ⅳ	2.11	121.37	166.03
Ⅴ	1.98	57.80	327.15

4. Design and calculation of transmission components

(1) Design and calculation of high-speed stage gear.

Chapter 20 Drawing Examples and Cases

1) Basic parameters of gear.

(a) Small and big gears of high-speed stage.

Small gear of high-speed stage:

Material: 40Cr, tempering.

Rigidity: 280HBS.

Number of teeth: $z_1 = 24$.

Big gear of high-speed stage:

Material: 45 steel, tempering.

Rigidity: 240HBS.

Number of teeth: $z_2 = 24 \times 3.9 = 93.6$, so $z_2 = 93$.

(b) Precision grade of the gear.

Choose grade 8.

2) Calculation of the contact strength of gear.

Design by the tooth surface contact strength, according to the formula:

$$d_{1t} \geqslant \sqrt[3]{\frac{2K_{Ht}T_1}{\varphi_d} \times \frac{u+1}{u} \times \left(\frac{Z_H Z_E Z_\varepsilon Z_\beta}{[\sigma_H]}\right)^2}$$

(a) Determine the variables in the formula.

I. Choose load factor: $K_{Ht} = 1.3$.

II. Choose zone factor: $Z_H = 2.43$.

III. According to the required figure in reference [7]:

$$\alpha_t = \arctan(\tan\alpha_n / \cos\beta) = \arctan(\tan 20°/\cos 15°) = 20.647°$$

$$\alpha_{at1} = \arccos(z_1 \cos\alpha_t / (z_1 + 2h_a \times \cos\beta)) =$$

$$\arccos(24 \times \cos 20.647°/(24 + 2 \times 1 \times \cos 15°)) = 29.996°$$

$$\alpha_{at2} = \arccos(z_2 \cos\alpha_t / (z_2 + 2h_a \times \cos\beta)) =$$

$$\arccos(93 \times \cos 20.647°/(93 + 2 \times 1 \times \cos 15°)) = 23.547°$$

$$\varepsilon_\alpha = [z_1(\tan\alpha_{at1} - \tan\alpha_{at}) + z_2(\tan\alpha_{at2} - \tan\alpha_{at})]/2\pi =$$

$$[24 \times (\tan 29.996° - \tan 20.647°) + 93 \times (\tan 23.547° - \tan 20.647°)]/2\pi = 1.639$$

$$\varepsilon_\beta = \varphi_d z_1 \tan\beta / \pi = 24\tan 15°/\pi = 2.047$$

$$Z_\varepsilon = \sqrt{\frac{4-\varepsilon_\alpha}{3}(1-\varepsilon_\beta)\frac{\varepsilon_\beta}{\varepsilon_\alpha}} =$$

$$\sqrt{\frac{4-1.639}{3}(1-2.047)\frac{2.047}{1.639}} = 0.652$$

$$Z_\beta = \sqrt{\cos\beta} = \sqrt{\cos 15°} = 0.983$$

IV. Torque of the small gear:

$$T_1 = 15.67 \text{ N} \cdot \text{m}$$

V. According to the required table in reference [7], the tooth width factor: $\varphi_d = 1$.

VI. The elastic influence coefficient: $Z_E = 189.8 \text{MPa}^{\frac{1}{2}}$.

Ⅶ. The contact strength limit of the small gear: $\sigma_{Hlim\,1} = 600$ MPa.
The contact strength limit of the big gear: $\sigma_{Hlim\,2} = 550$ MPa.
Ⅷ. Calculate the number of stress cycles by the formula:
$$N_1 = 60n_1 j L_h = 60 \times 1,420 \times 1 \times (2 \times 8 \times 300 \times 7) = 2.86 \times 10^9 \text{ h}$$
$$N_2 = 2.86 \times 10^9 / 3.9 = 7.33 \times 10^8 \text{ h}$$

Ⅸ. Calculate the contact fatigue allowable stress.
Safety factor: $S = 1$, by the required formula:
$$[\sigma_H] = \frac{K_{HN1} \sigma_{Hlim\,1}}{S} = 0.88 \times 600 = 534 \text{ MPa}$$
$$[\sigma_H] = \frac{K_{HN2} \sigma_{Hlim\,2}}{S} = 0.8 \times 550 = 489.5 \text{ MPa}$$

Take a small value
$$[\sigma_H] = [\sigma_H] = 489.5 \text{ MPa}$$

Here the average value should be taken.

(b) Calculate.

Ⅰ. Calculate the pitch circle diameter of the small gear
$$d_{1t} \geqslant \sqrt[3]{\frac{2K_{Ht} T_1}{\varphi_d} \times \frac{u+1}{u} \times \left(\frac{Z_H Z_E Z_\epsilon Z_\beta}{[\sigma_H]}\right)^2} =$$
$$\sqrt[3]{\frac{2 \times 1.3 \times 15.67 \times 10^3}{1} \times \frac{3.9+1}{3.9} \times \left(\frac{2.43 \times 189.8 \times 0.652 \times 0.983}{489.5}\right)^2} = 26.527 \text{ mm}$$

Ⅱ. Calculate the circular velocity.
$$v = \frac{3.14 \times 26.527 \times 1,420}{60 \times 1,000} = 1.97 \text{ m/s}$$

Ⅲ. Calculation of tooth width and module.
$$b = \varphi_d \times d_{1t} = 34.154, \quad m_n = \frac{d_1 \cos \beta}{Z_1} = \frac{34.755 \times \cos 15°}{24} = 1.28 \text{ mm}$$

Ⅳ. Calculation of the load factor.
$$K_{H\alpha} = K_{F\alpha} = 1.4, \quad K_A = 1$$
$$K_v = 1.1$$
$$K = 1.466 \text{ (From interpolation)}$$
$$h = 2.25 m_n = 2.25 \times 1.5 = 3.375 \text{ mm}, \quad \frac{b}{h} = b/h = \frac{37.436}{3.375} = 11.092$$

So we get: $K_{F\beta} = 1.41$.
So the load factor:
$$K_H = K_A K_v K_{H\alpha} K_{H\beta} = 1 \times 1.1 \times 1.4 \times 1.446 = 2.23$$

Ⅴ. Adjust the pitch circle diameter by the actual load factor:
$$d_1 = d_{1t} \sqrt[3]{\frac{K_H}{K_{Ht}}} = 26.527 \times \sqrt[3]{\frac{2.23}{1.3}} = 31.775 \text{ mm}$$

Ⅵ. Calculate the module:

Chapter 20 Drawing Examples and Cases

$$m_n = \frac{d_1 \cos \beta}{Z_1} = \frac{31.775 \times \cos 15°}{24} = 1.28 \text{ mm, so } m_n = 1.5 \text{ mm}$$

(c) Determine the final dimensions.

I. Calculate the distance between the centers of the two gears.

$$a = \frac{(z_1 + z_2)m_n}{2\cos \beta} = \frac{(24+93) \times 1.5}{2 \times \cos 15°} = 90.845 \text{ mm}$$

Round the center distance to 90 mm.

II. Revise the helix angle by the rounded centre distance.

$$\beta = \arccos \frac{(z_1 + z_2)m_n}{2a} = \arccos \frac{(24+93) \times 1.5}{2 \times 90} = 12.839° (12°50'18'')$$

III. Calculate pitch circle diameter of the two gears.

$$d_1 = \frac{z_1 m_n}{\cos \beta} = \frac{24 \times 1.5}{\cos 12.839°} = 36.923 \text{ mm},$$

$$d_2 = \frac{z_2 m_n}{\cos \beta} = \frac{93 \times 1.5}{\cos 12.839°} = 143.077 \text{ mm}$$

IV. Calculate the width of the gear.

$$B = \varphi_d d_1 = 1 \times 36.923 \text{ mm} = 36.923 \text{ mm}$$

Then $B_2 = 38$ mm, $B_1 = 44$ mm.

3) Checking of the bending strength of gear.

Calculate by the formula:

$$\sigma_F = \frac{2K_F T_1 Y_{Fa1} Y_{Sa1} Y_\varepsilon Y_\beta \cos^2 \beta}{\varphi_d m_n z_1} \leqslant [\sigma_F]$$

Determine the values in the formula.

Calculate load factor:

$$K_F = K_A K_V K_{F\alpha} K_{F\beta} = 1 \times 1.1 \times 1.4 \times 1.41 = 2.17$$

According to the overlap contact ratio: $\varepsilon_\beta = 1.741$, we get: $Y_\beta = 0.685$.

Calculate equivalent number of teeth:

$$z_{v1} = z_1/\cos^3 \beta = 24/\cos^3 12.839° = 25.894$$

$$z_{v2} = z_2/\cos^3 \beta = 93/\cos^3 12.839° = 100.339$$

Choose form factor from table in reference [7], $Y_{Fa1} = 2.63$ $Y_{Fa2} = 2.20$.

Choose stress correction factor from table in reference [7], $Y_{sa1} = 1.60$, $Y_{sa2} = 1.80$.

Choose the bending fatigue limit from figure in reference [7].

$$\sigma_{Flim\,1} = 500 \text{ MPa}, \quad \sigma_{Flim\,2} = 380 \text{ MPa}$$

Choose bending fatigue life factor from figure in reference [7]:

$$K_{FN1} = 0.82, \quad K_{FN2} = 0.87$$

Calculate the allowable bending stress fatigue, choose bending fatigue safety factor $S = 1.4$.

$$[\sigma_F]_1 = \frac{K_{FN1} \sigma_{Flim\,1}}{S} = \frac{0.82 \times 500}{1.4} = 293 \text{ MPa}$$

$$[\sigma_F]_2 = \frac{K_{FN2}\sigma_{Flim\,2}}{S} = \frac{0.87 \times 380}{1.4} = 236 \text{ MPa}$$

Calculate.

$$\sigma_{F1} = \frac{2K_F T_1 Y_{Fa1} Y_{Sa1} Y_\varepsilon Y_\beta \cos^2\beta}{\varphi_d m_n z_1} =$$

$$\frac{2 \times 2.171 \times 15{,}670 \times 2.63 \times 1.60 \times 0.685 \times 0.814 \times \cos^2 12.839°}{1^2 \times 1.5^3 \times 24^2} = 78.06 \leqslant [\sigma_F]_1$$

$$\sigma_{F2} = \frac{2K_F T_1 Y_{Fa2} Y_{Sa2} Y_\varepsilon Y_\beta \cos^2\beta}{\varphi_d m_n z_1} =$$

$$\frac{2 \times 2.171 \times 15{,}670 \times 2.2 \times 1.80 \times 0.685 \times 0.814 \times \cos^2 12.839°}{1^2 \times 1.5^3 \times 24^2} = 73.44 \leqslant [\sigma_F]_2$$

(2) Design and calculation of low-speed stage gear.

1) Basic parameters of gear.

(a) Gears of low-speed stage.

Small gear of low-speed stage:

Material: 40Cr, tempering.

Rigidity: 280HBS.

Number of teeth: $z_3 = 25$.

Big gear of low-speed stage:

Material: 45 steel, tempering.

Rigidity: 240HBS.

Number of teeth: $z_4 = 25 \times i_{34} = 25 \times 3 = 75$, so $z_4 = 76$.

(b) Accuracy grade of the gear.

Choose grade 8.

2) Calculation of the contact strength of gear.

Design by the tooth surface contact strength, according to the formula:

$$d_{3t} \geqslant \sqrt[3]{\frac{2K_{Ht}T_2}{\varphi_d} \times \frac{u+1}{u} \times \left(\frac{Z_H Z_E Z_\varepsilon Z_\beta}{[\sigma_H]}\right)^2}$$

After determining the variables in the formula, calculate the pitch circle diameter of the small gear:

$$d_{3t} \geqslant \sqrt[3]{\frac{2K_{Ht}T_2}{\varphi_d} \times \frac{u+1}{u} \times \left(\frac{Z_H Z_E Z_\varepsilon Z_\beta}{[\sigma_H]}\right)^2} =$$

$$\sqrt[3]{\frac{2 \times 1.3 \times 58.75 \times 10^3}{1} \times \frac{3+1}{3} \times \left(\frac{2.43 \times 189.8 \times 0.643 \times 0.983}{500.5}\right)^2} = 41.034 \text{ mm}$$

Calculate the circular velocity:

$$v = \frac{3.14 \times 41.034 \times 364.10}{60 \times 1{,}000} = 0.782 \text{ m/s}$$

Calculation of tooth width and module:

$$b = \varphi_d \times d_{3t} = 41.034, \quad m_n = \frac{d_3 \cos\beta}{z_1} = \frac{48.375 \times \cos 15}{25} = 1.869 \text{ mm}$$

Chapter 20 Drawing Examples and Cases

$$\varepsilon_\beta = \varphi_d z_1 \tan\beta / 3.14 = 1 \times 25 \times \tan 15.866° / 3.14 = 2.262$$

Calculation of the load factor:

$$K_{H\alpha} = K_{F\alpha} = 1.4, \quad K_A = 1$$

$$K_v = 1.1$$

$$K_{H\beta} = 1.450 \text{ (From interpolation)}$$

$$h = 2.25 m_n = 2.25 \times 2 = 4.5 \text{ mm}, \quad \frac{b}{h} = b/h = \frac{41.034}{4.5} = 9.119$$

We get: $K_{F\beta} = 1.39$.

So the load factor:

$$K_H = K_A K_v K_{H\alpha} K_{H\beta} = 1 \times 1.1 \times 1.4 \times 1.450 = 2.13$$

Adjust the pitch circle diameter by the actual load factor:

$$d_3 = d_{3t} \sqrt[3]{\frac{K_H}{K_{Ht}}} = 41.034 \times \sqrt[3]{\frac{2.13}{1.3}} = 48.375 \text{ mm}$$

Calculate the module:

$$m_n = \frac{d_3 \cos\beta}{Z_3} = \frac{48.375 \times \cos 15°}{25} = 1.869 \text{ mm, so } m_n = 1.5 \text{ mm}$$

Calculate the distance between the centers of the two gears:

$$a = \frac{(z_3 + z_4) m_n}{2 \cos\beta} = \frac{(25 + 76) \times 2}{2 \times \cos 15°} = 104.563 \text{ mm}$$

Round the center distance to 105 mm.

Revise the helix angle by the rounded centre distance:

$$\beta = \arccos \frac{(z_3 + z_4) m_n}{2a} = \arccos \frac{(25 + 76) \times 2}{2 \times 105} = 15.866° (15°51'56'')$$

Calculate pitch circle diameter of the two gears:

$$d_3 = \frac{z_3 m_n}{\cos\beta} = \frac{25 \times 2}{\cos 15.866°} = 51.980 \text{ mm}, \quad d_4 = \frac{z_4 m_n}{\cos\beta} = \frac{76 \times 2}{\cos 15.866°} = 158.020 \text{ mm}$$

Calculate the width of the gear:

$$B = \varphi_d d_3 = 1 \times 48.375 = 48.375 \text{ mm}$$

Then $B_4 = 48$ mm, $B_3 = 54$ mm.

3) Checking of the bending strength of gear.

Calculate by the formula:

$$\sigma_F = \frac{2 K_F T_2 Y_{Fa1} Y_{Sa1} Y_\varepsilon Y_\beta \cos^2\beta}{\varphi_d m_n z_3} \leqslant [\sigma_F]$$

Determine the values in the formula.

The load factor

$$K_F = K_A K_v K_{F\alpha} K_{F\beta} = 1 \times 1.05 \times 1.4 \times 1.39 = 2.043$$

According to the overlap contact ratio: $\varepsilon_\beta = 2.262$, we get: $Y_\beta = 0.680$.

Equivalent number of teeth:

$$z_{v3} = z_3/\cos^3\beta = 25/\cos^3 15.866 = 28.090$$
$$z_{v4} = z_4/\cos^3\beta = 76/\cos^3 15.866 = 58.392$$

Choose form factor, $Y_{Fa1} = 2.58$, $Y_{Fa2} = 2.23$.

Choose stress correction factor, $Y_{Sa1} = 1.62$, $Y_{Sa2} = 1.75$.

Choose the bending fatigue limit:
$$\sigma_{Flim\,3} = 500 \text{ MPa}, \quad \sigma_{Flim\,4} = 380 \text{ MPa}$$

Choose bending fatigue life factor:
$$K_{FN_1} = 0.87, \quad K_{FN_2} = 0.90$$

Calculate the allowable bending stress fatigue, choose bending fatigue safety factor $S = 1.4$.

$$[\sigma_F]_3 = \frac{K_{FN1}\sigma_{Flim\,1}}{S} = \frac{0.87 \times 500}{1.4} = 311 \text{ MPa}$$

$$[\sigma_F]_4 = \frac{K_{FN2}\sigma_{Flim\,2}}{S} = \frac{0.9 \times 380}{1.4} = 244 \text{ MPa}$$

Calculate.

$$\sigma_{F3} = \frac{2K_F T_2 Y_{Fa1} Y_{Sa1} Y_\epsilon Y_\beta \cos^2\beta}{\varphi_d m_n z_3} =$$

$$\frac{2 \times 2.043 \times 58,750 \times 2.58 \times 1.62 \times 0.680 \times 0.701 \cos^2 15.866°}{1 \times 2^3 \, 25^2} = 88.50 \leqslant [\sigma_F]_3$$

$$\sigma_{F4} = \frac{2K_F T_2 Y_{Fa1} Y_{Sa1} Y_\epsilon Y_\beta \cos^2\beta}{\varphi_d m_n z_3} =$$

$$\frac{2 \times 2.043 \times 58,750 \times 2.23 \times 1.75 \times 0.680 \times 0.701 \cos^2 15.866°}{1 \times 2^3 \, 25^2} = 82.64 \leqslant [\sigma_F]_4$$

(3) Design and calculation of the bevel gear.

Choose numbers of teeth: $z_1 = 19$, $z_2 = i_{bevel} \times z_1 = 2.1 \times 19 = 39.9$, so $z_2 = 40$.

1) Basic parameters of gear.

Small gear:

Material: 40Cr, tempering.

Rigidity: 280HBS.

Big gear:

Material: 45 steel, tempering.

Rigidity: 240HBS.

Accuracy grade of the gear: grade 8.

2) Calculation of the bending strength of gear.

According to the formula:

$$m_t \geqslant \sqrt[3]{\frac{K_{Ft} T_4}{\varphi_R (1 - 0.5\varphi_R)^2 Z_5 \sqrt{i_{bevel}^2 + 1}} \left(\frac{Y_{Fa} Y_{Sa}}{[\sigma_F]}\right)}$$

$$K_{Ft} = 1.3$$

$$T_4 = 166.03 \text{ N} \cdot \text{mm}$$

Chapter 20 Drawing Examples and Cases

$$\delta_5 = \arctan(1/i_{\text{bevel}}) = \arctan(1/2.1) = 25.463°$$
$$\delta_6 = 90° - 25.463° = 64.587°$$
$$z_{v5} = z_5/\cos\delta_5 = 19/\cos 25.463° = 21.044$$
$$z_{v6} = z_6/\cos\delta_6 = 40/\cos 64.587° = 93.210$$
$$Y_{\text{Fa5}} = 2.79, \quad Y_{\text{Fa6}} = 2.20$$
$$Y_{\text{Sa5}} = 1.56, \quad Y_{\text{Sa6}} = 1.81$$
$$\sigma_{\text{Flim 5}} = 500 \text{ MPa}$$
$$\sigma_{\text{Flim 6}} = 380 \text{ MPa}$$
$$N_5 = 60 n_4 j L_h = 60 \times 121.37 \times 1 \times (2 \times 8 \times 300 \times 7) = 2.45 \times 10^8$$
$$N_6 = 2.45 \times 10^8 / 2.1 = 1.17 \times 10^8$$
$$K_{\text{FN5}} = 0.88, \quad K_{\text{FN6}} = 0.90$$
$$S = 1.4$$
$$[\sigma_F] = \frac{K_{\text{FN5}} \sigma_{\text{Flim 5}}}{S} = \frac{0.88 \times 500}{1.4} = 314.29 \text{ MPa}$$
$$[\sigma_F] = \frac{K_{\text{FN6}} \sigma_{\text{Flim 6}}}{S} = \frac{0.90 \times 380}{1.4} = 244.29 \text{ MPa}$$

Calculate $\dfrac{Y_{\text{Fa}} Y_{\text{Sa}}}{[\sigma_F]}$.

$$\frac{Y_{\text{Fa5}} F_{\text{Sa5}}}{[\sigma_F]_5} = \frac{2.79 \times 1.56}{314.29} = 0.013,8, \quad \frac{Y_{\text{Fa6}} F_{\text{Sa6}}}{[\sigma_F]_6} = \frac{2.20 \times 1.81}{244.29} = 0.016,3$$

So $$m_t \geq \sqrt[3]{\frac{1.3 \times 166,030}{0.3(1 - 0.5 \times 0.3)^2 \times 19^2 \sqrt{2.1^2 + 1}} \times 0.016,3} = 2.684 \text{ mm}$$

Calculate the circular velocity.
$$d_5 = m_t \times z_5 = 2.684 \times 19 = 50.996$$
$$d_{m5} = d_5 \times (1 - 0.5 \times \varphi_R) = 50.996 \times (1 - 0.5 \times 0.3) = 43.347$$
$$v_m = \frac{\pi d_{m5} n_4}{60 \times 1,000} = \frac{\pi \times 43.347 \times 121.37}{60 \times 1,000} = 0.275$$

The tooth width
$$b = \varphi_R \times d_5 \sqrt{i_0 + 1/2} = 0.3 \times 50.996 \sqrt{2.1^2 + 1/2} = 33.90 \text{ mm}$$

Calculation of the load factor.
$$K_v = 1.0$$
$$K_{F\beta} = 1.23$$

So the load factor:
$$K = K_A K_v K_{F\alpha} K_{F\beta} = 1 \times 1.0 \times 1 \times 1.23 = 1.23$$

Adjust the pitch circle diameter by the actual load factor:
$$m = m_t \sqrt[3]{\frac{K_F}{K_{Ft}}} = 2.684 \times \sqrt[3]{\frac{1.23}{1.3}} = 2.635 \text{ mm}$$

Calculate the module.

$$m = 3 \text{ mm}$$
$$d_5 = m \times z_5 = 3 \times 17 = 51$$
$$d_6 = m \times z_6 = 3 \times 36 = 108$$
$$\delta_5 = \arctan(1/u) = \arctan(17/36) = 25.278°(25°16'39'')$$
$$\delta_6 = 90° - 25.278° = 64.722°(64°43'21'')$$

5. Design and calculation of the shaft.

(1) Design of the high-speed shaft.

1) Structural design of the high-speed shaft.

(a) Estimate the minimum diameter.

According to the selected material: $A_0 = 112$.

We can calculate the minimum diameter of the shaft:

$$d_{\min} = A_0 \sqrt[3]{\frac{P_1}{n_1}} = 112 \sqrt[3]{\frac{2.33}{1,420}} = 13.2 \text{ mm}$$

Because there are two keys, it should be increased by 10%–15%.

$d_{\min} = 13.2 + 13.2 \times (10\% - 15\%) \text{ mm} = 13.2 + (1.32 - 1.98) = 14.52 - 15.18 \text{ mm}$

The segment of the minimum diameter of the shaft is connecting with the coupling.

(b) Determine the dimensions of the shaft.

According to the requirements of axial location, determine the shaft diameter and length of each segment (see Fig. 20-19).

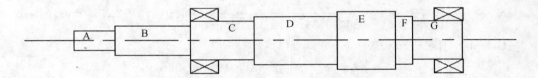

Fig. 20-19 Dimensions of high-speed shaft

Ⅰ. Determine the diameters.

Segment A: $d_1 = 18$ mm. The diameter of the half coupling is considered. Here the output shaft diameter of the motor should be considered, too.

Segment B: $d_2 = 20$ mm. It is determined according to the standard of sealing and the shoulder height.

Segment C: $d_3 = 25$ mm. To fit the bearing, it is equal to the bearing inner diameter.

Segment D: $d_4 = 29$ mm. Locating shoulder height here is considered and a gear shaft is designed.

Segment E: $d_5 = 36.923$ mm. The gear teeth is located here.

Segment F: $d_6 = 29$ mm. The shoulder height is considered.

Segment G: $d_7 = d_3 = 25$ mm.

Ⅱ. Determine the length of each segment.

Segment A: $L_1=27$ mm. The length of the half coupling is considered.

Segment B: $L_2=50$ mm. The width of bearing cover and the distance between bearing cover and the coupling are considered.

Segment C: $L_3=B+$oil baffle width$=15+19=34$mm. The bearing and oil baffle are considered.

Segment D: $L_4=67$ mm.

Segment E: $L_5=b_1=44$ mm. The width of the gear is 44 mm.

Segment F: $L_6=10$ mm.

Segment G: $L_7=30$ mm. The width of the bearing is considered.

2) Strength checking of the high-speed shaft.

(a) Stress analysis of the shaft.

I. The force analysis of the high speed shaft is shown in Fig. 20-20.

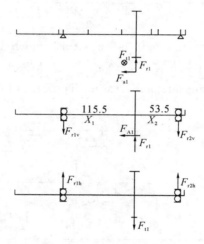

Fig. 20-20 Force analysis of the high speed shaft

$$F_{t1}=\frac{2T_1}{d_1}=\frac{2\times 15.67\times 10^3}{36.923}=848.8 \text{ N}$$

$$F_{r1}=\frac{\tan\alpha_n}{\cos\beta}\cdot F_{t1}=\frac{\tan 20°}{\cos 12.839°}\times 848.8=316.9 \text{ N}$$

$$F_{a1}=F_{t1}\tan\beta=848.8\times\tan 12.839°=193.5 \text{ N}$$

II. Calculate the supporting forces:

$$X_1=15/2+19+67+44/2=115.5 \text{ mm}$$

$$X_2=44/2+10+15+12/2=54.5 \text{ mm}$$

On the horizontal plane:

$$F_{r2h}=\frac{115.5}{115.5+54.5}F_{t1}=848.8\times\frac{115.5}{115.5+54.5}=576.7 \text{ N}$$

$$F_{r1h}=F_{t1}-F_{r2h}=848.8-576.7=272.1 \text{ N}$$

On the vertical plane

$$F_{r2v} = \frac{F_{r1} \times 115.5 + \dfrac{F_{a1} 36.932}{2}}{115.5 + 54.5} = \frac{316.9 \times 115.5 + \dfrac{193.5 \times 36.932}{2}}{115.5 + 54.5} = 194.3 \text{ N}$$

$$F_{r1v} = F_{r1} - F_{r2v} = 316.9 - 194.3 = 122.6 \text{ N}$$

(b) Calculation according to the condition of bending and torsion.

$$M_h = F_{r1h} \times X_1 = 272.1 \times 115.5 = 31,427.6 \text{ N}$$
$$M_{v1} = -F_{r1v} \times X_1 = -122.6 \times 115.5 = -14,160.3 \text{ N}$$
$$M_{v2} = F_{r2v} \times X_2 = 194.3 \times 54.5 = 10,589.4 \text{ N}$$
$$M_1 = \sqrt{M_h^2 + M_{v1}^2} = \sqrt{31,427.6^2 + (-14,160.3)^2} = 34,470.4 \text{ N}$$
$$M_2 = \sqrt{M_h^2 + M_{v2}^2} = \sqrt{31,427.6^2 + 10,589.4^2} = 33,163.7 \text{ N}$$
$$T_1 = 15.67 \text{ N} \cdot \text{m}$$

We choose $a = 0.6$. Calculate the stress of the shaft:

$$\sigma_{ca} = \frac{\sqrt{M^2 + (\alpha T_2)^2}}{W} = \frac{\sqrt{34,470.4^2 + (0.6 \times 15,670)^2}}{3.14 \times 36.923^3 / 32} = 7.15 \text{ MPa}$$

The selected material of the shaft is 45 steel, which is hardened and tempered. According to the table in reference [7], get $[\sigma_{-1}] = 60$ MPa, $\sigma_{ca} < [\sigma_{-1}]$. So it meets the requirements. Draw the bending moment, torquediagram as follows in Fig. 20-21.

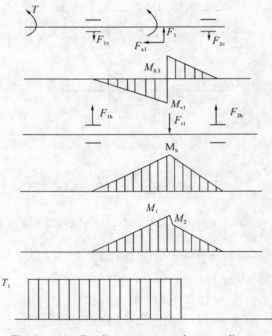

Fig. 20-21 Bending moment and torque diagram

3) Fatigue strength calculation for the high speed shaft (canceled).

The procedure and the method is similar to 20.2.

(2) Design of the middle shaft.

(3) Design of the low-speed shaft.

Chapter 20 Drawing Examples and Cases

6. Selection and calculation of bearing life

(1) Calculation of bearings on the high-speed shaft

Choose angular contact ball bearing 7205AC.

Force analysis of bearings on the high-speed shaft is shown in Fig. 20 - 22.

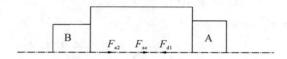

Fig. 20 - 22   Force analysis of high-speed shaft bearings

Its parameter:

$$d \times D \times B = 25 \text{ mm} \times 52 \text{ mm} \times 15 \text{ mm}, \quad C_r = 12.2 \text{ kN},$$

$$F_{r_1} = \sqrt{F_{r1v} + F_{r1h}} = \sqrt{112.6^2 + 272.1^2} = 298.4 \text{ N},$$

$$F_{r_2} = \sqrt{F_{r2v} + F_{r2h}} = \sqrt{194.3^2 + 576.7^2} = 608.6 \text{ N}$$

$$F_{d_1} = 0.68 F_{r1} = 0.68 \times 298.4 = 202.9 \text{ N}$$

$$F_{d_2} = 0.68 F_{r2} = 0.68 \times 608.6 = 413.8 \text{ N}$$

So

$$F_{ae} = F_{a1} = 193.5 \text{ N}$$

$$F_{d_2} + F_{ae} = 193.5 + 413.8 = 607.3 \text{ N}$$

$$F_{d_1} = 202.9 \text{ N}$$

$$F_{d_2} + F_{ae} = 193.5 + 413.8 = 607.3 \text{ N} > F_{d_1} = 202.9 \text{ N}$$

Pressed end 1:

$$F_{ca1} = F_{d_2} + F_{ae} = 607.3 \text{ N}$$

loosened end 2:

$$F_{ca2} = F_{d_2} = 413.8 \text{ N}$$

$$\frac{F_{ca1}}{F_{r1}} = \frac{607.3}{294.9} = 2.059 > 0.68$$

$$\frac{F_{ca2}}{F_{r2}} = \frac{413.8}{612.1} = 0.676 < 0.68$$

Equivalent dynamic load of bearing P_1 and P_2.

$$X_1 = 0.41, \quad Y_1 = 0.87$$

$$X_2 = 1, \quad Y_2 = 0$$

$$P_1 = f_d(XF_r + YF_{ca}) = 1 \times (0.41 \times 298.4 + 0.87 \times 607.3) = 650.695 \text{ N}$$

$$P_2 = f_d(XF_r + YF_{ca}) = 1 \times (1 \times 608.6 + 0 \times 413.8) = 608.6 \text{ N}$$

Because

$$P_1 > P_2$$

$$L_h = \frac{10^6}{60 n_1}\left(\frac{C}{P}\right)^\varepsilon = \frac{10^6}{60 \times 1,420} \times \left(\frac{12.2 \times 1,000}{650.695}\right)^3 = 77,358 \text{ h} > 33,600 \text{ h}$$

It meets the requirements.

(2) Calculation of bearings on the middle shaft.

(3) Calculation of bearings on the low-speed shaft.

They both meet the requirements.

7. Selection and calculation of key

(1) The key on the high-speed shaft.

1) Key for the flexible coupling.
$$b \times h \times L = 6 \text{ mm} \times 6 \text{ mm} \times 22 \text{ mm}$$

2) Calculation of key.
$$l = L - b = 22 - 6 = 16 \text{ mm},$$
$$\sigma_p = 4,000 \times T_1/(h \times l \times d) = 4,000 \times 15.67/(6 \times 6 \times 18) = 36.3 < 120 \text{ MPa}$$

So it meets the requirements.

(2) The key on the middle shaft.

(3) The key on the low-speed shaft.

They both meet the requirements.

8. Selection of coupling

(1) Coupling on the input shaft.
$$K_A = 1.5, \quad T_{ca} = K_A T_3 = 1.5 \times 15.67 = 23.505 \text{ N} \cdot \text{m}$$

TL3 is selected.

Driving end $d_2 = 16$ mm, $L = 30$ mm.

Driven $d_2 = 18$ mm, $L = 30$ mm.

Nominal torque $T_n = 31.5 \text{ N} \cdot \text{m} > 23.505 \text{ N} \cdot \text{m}$.

Allowable rotating speed $[n] = 4,700 > 1,420$ r/min.

So it meets the requirements. *But here the output shaft diameter of the selected motor is not taken into account.*

(2) Coupling on the output shaft.
$$K_A = 1.5, \quad T_{ca} = K_A T_3 = 1.5 \times 169.17 = 253.755 \text{ N} \cdot \text{m}$$

HL2 is selected.

Driving end $d_2 = 30$ mm, $L = 60$ mm.

Driven $d_2 = 35$ mm, $L = 60$ mm

Nominal torque $T_n = 315 \text{ N} \cdot \text{m} > 253.755 \text{ N} \cdot \text{m}$

Allowable rotating speed $[n] = 5,600 > 121.37$ r/min.

So it meets the requirements.

9. Lubrication and sealing

The information about lubrication and sealing should be mentioned here, and 20.2 can be referred to.

Fig. 20-23 to Fig. 20-25 show the assembly and some component drawings of this case.

Chapter 20 Drawing Examples and Cases

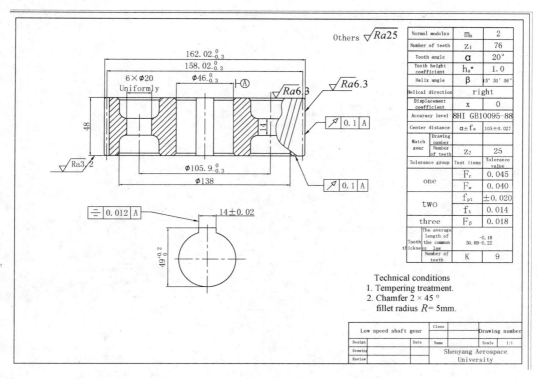

Fig. 20 – 23 Gear drawing

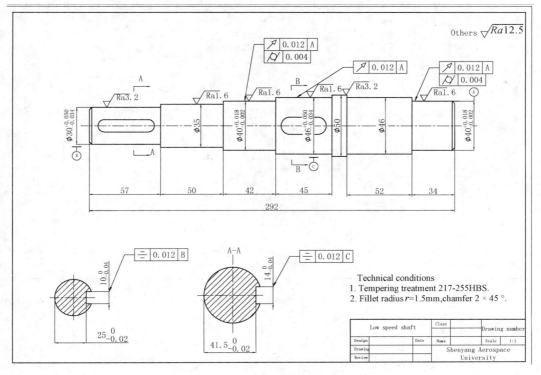

Fig. 20 – 24 Low-speed shaft drawing

· 165 ·

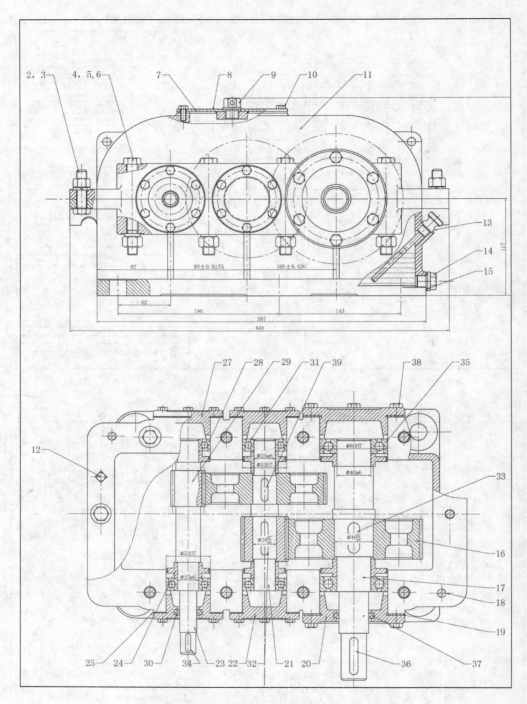

Fig. 20-25 Reducer assembly drawing

Chapter 20 Drawing Examples and Cases

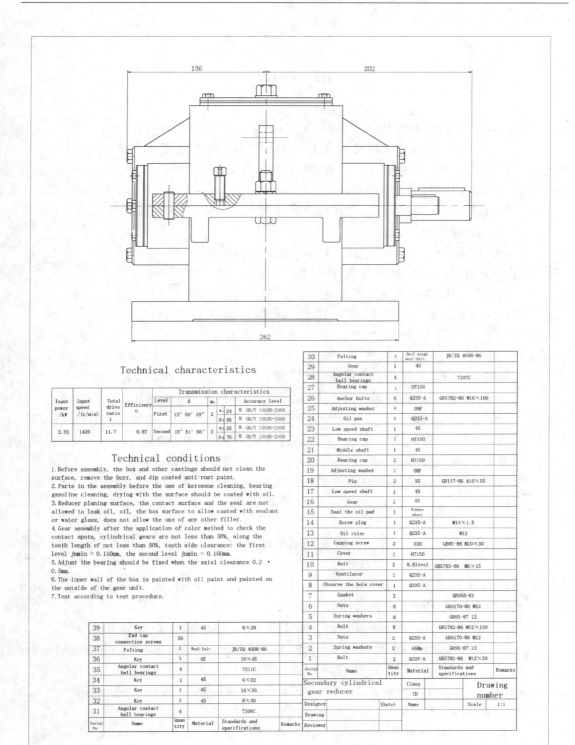

Continued Fig. 20-25 Reducer assembly drawing

Reference

[1] 濮良贵,纪名刚. 机械设计[M]. 8版. 北京:高等教育出版社,2006.

[2] 孔云鹏,田万禄,张祖立,等. 机械设计课程设计[M]. 沈阳:东北大学出版社,2000.

[3] 张锋,古乐. 机械设计课程设计手册[M]. 北京:高等教育出版社,2010.

[4] 吴宗泽,罗圣国. 机械设计课程设计手册[M]. 4版. 北京:高等教育出版社,2012.

[5] 张彦富,付求涯,王阿春,等. 几何量公差与测量技术基础(英文版)[M]. 北京:北京航空航天大学出版社,2015.

[6] 成大先. 机械设计手册[M]. 北京:化学工业出版社,2010.

[7] 濮良贵,陈国定,吴立言. 机械设计[M]. 9版. 北京:高等教育出版社,2013.

[8] 于惠力,冯新敏. 现代机械零部件设计手册[M]. 北京:机械工业出版社,2013.

[9] 金嘉琦. 几何量精度设计与检测[M]. 北京:机械工业出版社,2012.

[10] 王大康. 机械设计课程设计[M]. 北京:中国铁道出版社,2015.

[11] 冯立艳,李建工,陆玉. 机械设计课程设计[M]. 5版. 北京:机械工业出版社,2016.

[12] 孔云鹏,田万禄,张伟华,等. 机械设计课程设计[M]. 北京:科学出版社,2008.

[13] 寇尊权,王多. 机械设计课程设计[M]. 2版. 北京:机械工业出版社,2011.

[14] 《机械设计手册》编委会. 机械设计手册:齿轮传动[M]. 北京:机械工业出版社,2007.

[15] 陆玉. 机械设计课程设计[M]. 4版. 北京:机械工业出版社,2007.

[16] 李育锡. 机械设计课程设计[M]. 2版. 北京:高等教育出版社,2014.

[17] 韩晓娟. 机械设计课程设计指导手册[M] 北京:中国标准出版社,2008.

[18] Yang Ming-Zhong. Machinery Design[M]. Wuhan:Wuhan University of Technology Press,2004.

[19] Edwards K S,McKee R B. Fundamentals of Mechanical Component Design[M]. New York:McGraw-Hill,1991.

[20] Spotts M F,Shoup T E,Horn berger L E. Design of Machine Elements[M]. Upper Saddle Kiver:Prentice-Hall,Inc. ,2003.

[21] Dimarogoneas A D. Machine Design[M]. Hoboken:John Wiley & Sons,Inc. ,2001.

[22] Juvinall R C,Marshek K M. Fundamentals of Machine Component Design[M]. 3rd ed. Hoboken:John Wiley & Sons,Inc. ,2003.

[23] Smith E H. Mechanical Engineer's Reference Book[M]. 12th ed. Oxford:Butterworth-Heinemann,1998.

[24] Mott R L. Machine Elements in Mechanical Design[M]. 3rd ed. Upper Saddle River:Prentice-Hall,Inc. ,1999.

[25] Shigley J E. Mechanical Engineering Design[M]. 6th ed. New York:McGraw-Hill,2001.